Bringing Back the Bush

Eileen and Joan Bradley.

COURTESY OF JOHN FAIRFAX AND SONS LTD.

Bringing Back the BUSH

THE BRADLEY METHOD OF BUSH REGENERATION

JOAN BRADLEY

Edited by
Joan Larking, Audrey Lenning and Jean Walker

Dedicated to Eileen, my sister who died in February of 1976, but speaks with me through every line.

Published in 2002 by Reed New Holland
An imprint of New Holland Publishers
Sydney

Level 1, 178 Fox Valley Road, Wahroonga, NSW 2076, Australia

newhollandpublishers.com

First published by Lansdowne-Rigby 1988
Reprinted by Ure Smith Press 1991
Reprinted by Lansdowne Press Pty Ltd 1997
Reprinted by Reed New Holland 2002, 2017, 2024

A record of this book is held at the British Library and the National Library of Australia.

ISBN 9781876334895

Group Managing Director: Fiona Schultz
Cover Design: Nanette Backhouse
Cover Photograph: Ivy Hansen Photography
Typesetter: Savage Type, Brisbane
Printed in China

Keep up with New Holland Publishers:
NewHollandPublishers
@newhollandpublishers

ACKNOWLEDGEMENTS

On behalf of Eileen and Joan Bradley's family, I would like to pay tribute to the skill and dedication of Joan Larking, Audrey Lenning and Jean Walker who undertook the demanding task of completing and editing Joan's manuscript. Drawing on her suitcase full of notes, some neatly compiled, some in more disarray, they have, I believe, successfully retained Joan's style of writing and provided a clear and accurate account of those techniques and problems not covered in the unfinished text.

Joan Larking was trained by Joan and was undoubtedly her star pupil with ten years' very wide experience in the field. She taught the Bradley Method for a number of years and enjoyed almost daily contact with Joan discussing the book.

Audrey Lenning, a finance journalist, is also a trained regenerator who, like June Gram, worked with Eileen and Joan in Ashton Park. Both were on the Committee of the Mosman Parklands and Ashton Park Association which published the successful forerunner to this publication, *Bush Regeneration.* Without the support of the Association, the Bradley Method might never have got off the ground.

Jean Walker, landscape architect and co-author of *Designing Australian Bush Gardens* and *More About Bush Gardens,* worked closely with Joan in formulating her ideas since 1968. When the budget for photographs ran out, she provided drawings and additional photographs in two weeks. These illustrations have been wonderfully supplemented by Ian Walker's whimsical cartoons. Ian, an early volunteer regenerator, teaches the Bradley Method on the North Coast of New South Wales.

Neil Mudford, Chairman of the Total Environment Centre Management Committee, provided helpful advice at a critical stage of the book's production.

Joan would also wish me to recognise the contribution of all the regenerators who provided ideas and refined and invented techniques for dealing with new problems in the field.

To all the people who were Joan's sounding boards, our thanks.

Elizabeth Elenius

CONTENTS

INTRODUCTION

Many years have elapsed since the publication in 1967 of *Weeds and Their Control* and in 1971 of *Bush Regeneration;* years offrustration, gradual recognition, and finally acceptance of the principles and techniques developed and enunciated by Eileen and Joan Bradley. *Bringing Back the Bush* published now five years after Joan's untimely death and ten years after Eileen died, is the culmination of their work.

My earliest memories of 'the girls' as they were affectionately called by the family right up to their deaths, were of cheerful, extrovert and extremely interesting people. When I was eight my mother (their sister) used to read their letters from England. They were wonderful accounts of the field flowers and animals of the hedges, the people they met and the famous sights they visited. This was probably a turning point in their lives. Their mother had been a keen gardener and they had grown up with an awareness of plants, but it was their frequent visits to Virginia Waters and to the ancient woodlands of Burnham Beeches whilst overseas which stimulated what was to become for them both a lifelong interest in plant growth.

As I grew older, it was a great thrill to stay with them at Clifton Gardens. I was never bored. There was snorkelling in the little rock pool they had constructed below the house. On, walks in Ashton Park 'the girls' pointed out tiny wondrous plants clinging to the sandstone, pardelots' nests in the banks of the path and a myriad of other birds. Then of course, there were the antics of the blue wrens in the garden. Blue wrens were the subject of their five-year study which was published by the Royal Ornithological Union of Australia.

I remember sitting in the kitchen while Eileen and Joan, peering through binoculars, recorded their sightings of the birds which they had banded. They were greatly saddened by the gradual disappearance of their wrens through the introduction of garden sprays and predation by domestic animals.

Their enthusiasm for nature led to a wide and increasing involvement in conservation issues. Their commitment was catching and, in 1965, I found myself penning my very first letter to the editor, from which time I was hooked on conservation and I learned much from their broad knowledge of such issues as control burning and from their experience in political lobbying.

Passion would not be too strong a word to describe their interest in compost. Their compost heap was reverently called 'Sir Albert' after Sir Albert Howard who observed the benefits of compost in India between 1905 and 1924. Almost all garden rubbish, apart from onion weed and wandering jew, was chopped up with a cleaver and recycled in 'Sir Albert' for mulching their English cottage style garden.

Eileen and Joan had a great love of literature. Their library contained books on philosophy, the arts, history, exploration and travel (including diaries of early Australian explorers and naturalists), biographies and autobiographies, as well as plays and novels. Both were practical, but in different spheres. Eileen excelled in the domestic crafts of cooking, gardening and sewing; Joan, in her youth, raced with the Sydney Dinghy Club and skiied on the Main Range of the Australian Alps in the pioneer days of the sport. In addition, Joan was very skilled with her hands, making such items as chess sets, built-in cupboards, and, latterly, the toolboards described herein, for use in the field. She was also a brilliant black and white photographer.

Some of the story of bush regeneration is told by Joan in the first chapter of this book. Typically, it conveys only the excitement and thrill of the sisters' unfolding discoveries, not the disappointment and hurt experienced when the brilliance and simplicity of their regeneration techniques were spurned by ignorant and sometimes pigheaded individuals and bureaucracies.

I can remember their utter despair when, against the informed opposition of local conservation groups, the Forestry Commission prescribed burnt one hectare of Ashton Park and the anticipated spurt of weed growth occurred; their excitement at the strong growth of natives following the summer wildfire at Chowder Head; their good-humoured dismissal of the 'ratbag' tag given them by proponents of development, burning and crash through or crash weed control programmes; Joan's sadness, and joy, when official recognition finally arrived – sadness because Eileen died before recognition had been achieved, and joy because now many other bushland areas could be helped to return to their natural condition.

That sadness and joy is again experienced by those of us who were left behind after Joan's death to take on the challenge of finishing the book, a legacy we felt must not be lost.

The turning point for bush regeneration came in 1975 when the National Trust commissioned Joan, Toni May and their small team of regenerators to demonstrate their techniques in Blackwood Reserve, Beecroft. While regenerating Blackwood, Joan also proved to herself that the principles established in Hawkesbury sandstone bushland could also be applied to the moist schlerophyll woodland growing on the richer shale-derived soils and, ultimately, rainforest. With the support and sponsorship of the National Trust and, particularly, the management and promotion skills of Evelyn Hickey, the Trust's Regeneration Officer, the demand by local councils for the services of trained regenerators grew rapidly. I recall Joan saying, with a touch of irony, 'Now I'm being paid; I'm respectable'.

With demand for regenerators outstripping supply, a school was established to teach the Bradley method to conservationists keen to assist in bringing back their local bushland. Joan was commissioned to provide tuition and gradually that small band of previously unpaid workers grew – former pupils became teachers, and the Bradley Method is now being used throughout Australia and in some countries overseas.

There is a danger, however, that people may try to adapt the method and thus stray from the basic principles which MUST be adhered to. The publication of this book will enable those sympathetic to the bush to better understand and apply the Bradley Method of bush regeneration. Even with the success of bush regeneration, both as an ecologically and economically sound practice, there have been the knockers. It is particularly infuriating to me that Joan's writing and methods .have been labelled 'unscientific'. Joan was a qualified scientist, but she was much more than a conventional scientist. She was a true naturalist. Both she and Eileen had the rare ability to observe, to interpret and to extract from their observations simple general principles which can be explained, not by complex equations and esoteric language, but in simple English which everyone can understand. In Joan's words, 'As a very old-fashioned scientist and former chemist, I had a thorough grounding in what was then the simple scientific method of experiment and observation. Repeatability still remains for me the acid test. This method is repeatable anywhere as long as the three principles are followed'.

The value of this book lies as much in the way in which its message is delivered, as in the message itself – the scientific principles and techniques which have been established through trial, success and replication. Not for Joan the computer crutch or scientific catch-cries, but make no mistake, the development of the principles was conducted in the most scientifically rigorous manner.

Eileen and Joan Bradley were remarkable women who contributed much to nature conservation in New South Wales. Their enthusiasm for each project – the Superb Blue Wren study; the establishment of the Mosman Parklands and Ashton Park Association; the Working Kelpie Council; the cultivation of rare irises and liliums – was infectious and is greatly missed. Their legacy lies not just in this work but in the many bushland areas which have been (and will be) reclaimed for the enjoyment of future generations.

Elizabeth Elenius

HOW IT ALL BEGAN

My sister and I had for years been pulling up seedling weeds growing near the walking tracks in Ashton Park, and had looked despondently at the big ones scattered through the bush further in. We had always found these widespread invaders particularly offensive, and longed for the strength we believed was needed to cope with them. We felt that, because of their threat to the whole of the bush, these should be the first weeds to be destroyed, and were therefore delighted to see unsightly walls of tall lantana fall to the mattocks and brushhooks of the park staff.

We had never thought it possible that such very bad areas could be restored by anything other than this sort of clearing followed by replanting. The clearing was mostly confined to very heavy lantana infestations, where the few native seedlings that came up were quickly swamped by an explosion of assorted weeds, but in a few places work was extended into areas of mixed weeds and natives. Here, where they were not hopelessly outnumbered, the natives responded magnificently. Shrubs, despite disturbed roots and broken branches, put out new shoots, and seedlings of many species germinated along with the weeds.

With growing enthusiasm, we began to understand that there might be another way to fight the invaders. Given half a chance, the bush would fight back on its own behalf.

Nevertheless, when we wrote *Weeds and Their Control* in 1967, we had not realised just how resilient the native plants were. We believed, as we said then, that 'Ideally, the weed problem would best be solved by the complete removal of all exotic species, followed by further clearing of all weed regrowth and immediate replanting with advanced native seedlings.' We did realise that 'This would require an enormous staff, which is simply unavailable, and might moreover be defeated by adverse weather conditions.'

For these strictly common sense reasons, we recommended the essential strategy which I and many others are still using. We were, even then, using it ourselves, working our way through from lightly weed infested areas to heavily infested areas, but had still to test it on the areas of heaviest infestation.

By 1968, we had proved to our own satisfaction that, in one district at least, systematic hand weeding, carefully done, was a spectacular success. We made an addendum to *Weeds and Their Control,* with a map showing the stabilisation of half a hectare of bush, including part of a site, cleared by crude methods in 1965, which by 1967 had reverted to lantana well above our heads.

> The [park] staff worked for less than two days, grubbing out a large privet tree ... and burning uprooted lantana and slightly extending primary clearing. The remainder of the work was done by one woman in odd half hours throughout the year.

I was that one woman, then rising 52 years old. I may add that I was defeated by the 'slightly extended' clearing, which was done with mattocks. I have since learned that such sites are difficult to rid completely of weeds, but that it is less hard if they are given a few years in which to settle down.

> The work was intermittent but systematic: small areas were weeded in many places; time was given for weeds to regrow and seedlings to be found and removed, and for native plants to grow up and stabilise each area in turn. Great care was taken to cause minimum soil disturbance and a mulch of leaves and twigs was swept back over any soil exposed. Clearing was never allowed to outstrip regeneration, nor was any area allowed to stagnate for want of further clearing.

We calculated that, on the most liberal estimate, the time taken to clear the half hectare was equivalent to ten days' work by the three-man park staff. My sister, five years my senior, was no less successful, but she was working mainly on a dry ridge, where she was doing spot regeneration, and on Chowder Head, where we had no ground plan good enough to form the basis of a weed survey.

We did not overwork ourselves. My sister took the dog for a walk on most mornings and I did the same in the afternoons. On those walks we averaged, between the two of us, roughly three-quarters of an hour spent in actually pulling up weeds. We did not work in the rain, or harm the soil by puddling in it when it was still wet after heavy rainfall. Drought, too, stopped us as the ground grew too hard, and sometimes we felt less inclined to weed than to go for a good long walk.

Weeding in this way, we found that we had developed the basic technique needed to remove weeds in a way that favoured natives. Keen gardeners both, we knew the value of surface mulching, and the importance of the humus-rich upper soil layer. So we churned up mulch and soil as little as possible, and put everything back as nearly as we could in the right order.

In our weeding programme we had first pulled out the smaller weeds; then we had tried to remove the bigger weeds, loosening each root individually when they refused to budge. When this failed we used secateurs; and when roots and branches got too thick for secateurs, we used our garden pruning shears and finally the hatchet. At some stage, I took to using a sheath-knife. It was quite a surprise and most exhilarating to find that two women, no longer young, could quite easily handle the

Joan's photo of the walking track at Ashton Park; where it all began.

biggest lantana and bitou bush that our local reserves had to offer.

A weed can be defined as a plant out of place. In the native bush, exotics like the small wandering jew, right up to the large camphor laurel, are all growing out of place and are all classed as *weeds*. And we are dealing with weeds at all stages of their growth, from seedlings to mature trees. Some small shallow-rooted weeds can be removed simply by pulling; with the larger ones, the deep main roots and laterals may have to be traced out and removed individually, before the plant itself can be taken out.

We very soon discovered that it was impossible to dissect out roots of individual weeds in areas of good bush without interfering with the roots of natives. We knew that native plants hated to have their roots disturbed, and even without this knowledge we would have been most reluctant to damage the very plants that we wanted to preserve. So we traced the roots of the weeds until we could go no further without disturbing natives, cut them, covered the cut ends with soil, then mulched and awaited results.

Our reward was a joyous spurt of growth from the natives we had freed from competition – and no regrowth from the cut ends of the weed roots. We had found ourselves a new and fascinating sparetime occupation – bringing back the bush.

Weed control fails when you treat the work like farming or gardening – clearing away the plants that you do not want, and cultivating plants that you do. I have quite lost count of the times I have seen repeats of our 1964 Ashton Park clearing programme. More and more people are realising that there are foreign plants – weeds – in their local bush. More and more public authorities are trying to do something about it.

Others, meeting opposition to the razing of bush for various developments, defend themselves by saying that it is nothing but a mess of privet, lantana or morning glory. Members of conservation societies form working bees to display the bush behind the weed barrier, and to prove their willingness to work for it.

Most people, when clearing areas of bush, attack the weeds where they are most visible. Unhappily, this is exactly where they are the least vulnerable. Often, particularly when the work is done by voluntary labour or paid for by some sort of windfall grant, the area cleared is simply allowed to be reclaimed by the weeds. At best, it is kept under continuous care for an indefinite number of years. Various methods of weed control are used. Weeds are dug out or poisoned, which does no good to the few natives that have struggled up after the first onslaught. Adding mulch helps, but weeds grow up through it; also it disintegrates and has to be replaced. Eventually, volunteers lose heart and give up. Local councils weary of voting ratepayers' money for unproductive work.

Area after area which has been cleared is allowed to revert to weeds. The weeds, young and vigorous, take over the strip of bush next to the clearing which, because of the disturbance, has been left exposed and weakened. We are still waiting to see the reverse process – bush moving out and taking over successfully planted clearings.

Weed control succeeds when you bring back the bush – the strategy of favouring natives against weeds. To do this may involve a change of attitude. In our method of bush regeneration, it is the natives we are thinking about. We are concentrating, not on eradicating weeds, but on enabling native plants to grow, unhampered, in the environment that suits them best. This is a good exercise in applied ecology, encouraging the bush to control the invaders – for itself, yourself and for others in the future. Nothing you can do will speed this process beyond the natural growth rate of the native plants but you may do a lot to slow it down if you are not careful. Direct your mind always, not towards the slaughtered weeds, but towards growing natives.

In 1971, we realised that *Weeds and Their Control* was not positive enough in its approach, and I wrote *Bush Regeneration,* which was directed mainly towards amateur conservationists like ourselves. Explaining the reasoning behind our methods, I tried to convey the delight we felt in helping the bush to help itself. Much of the material in that book has been incorporated into the present volume which we hope will serve as a manual for future generations of concerned regenerators.

When the Mosman Parklands and Ashton Park Association was formed, and a bush regeneration team worked every Saturday morning, we began to teach in earnest. It was then that we found out how exceedingly skilful we had become at the work. We had not realised how much expertise we had developed, by trial and error over the years.

At first sight – even at first trial – our techniques may seem far too slow and too painstaking, especially when you contemplate a determined working party with brushhooks and mattocks, bashing down and grubbing out the weeds. They appear to be getting a lot of work done in a

Working from good areas to bad.

The challenge and the temptation. Experience will guide you. (Morning glory)

short time and enjoying the exercise too.

Compared with all that action, our weeding teams, tiptoeing about, equipped with absurdly small tools, snipping and probing and scratching, relaxed and unhurried, seem to be doing very little. But this seemingly leisurely approach is by no means as slow as it looks; there is so much that others have to do that our teams do not.

To begin with, we do not clear a site, amassing quantities of uprooted material which has to be carted away or burnt on the spot. We avoid double handling of uprooted weeds; we carry very little material out of the bush and we burn nothing. We do not propagate or buy costly native plants, dig holes for them, stake and water them, then tend them until they are established. Our native plants look after themselves. Follow-up, which is the second weeding session, takes a small fraction of the time spent on the first or primary weeding.

Our techniques, which save time by saving motions, may at first seem frustratingly slow, but do not become impatient. Practise on the smaller weeds until you have mastered the use of our highly selective, somewhat unconventional tools. It will not be too long before you can enjoy the surprise on people's faces when you show them how smoothly and easily you can take out a big weed and how little disturbance you create in doing it.

Bringing back bush involves two quite different kinds of time. The first is the time spent working. This includes travelling, primary weeding and follow-up. If the work is done by paid labour, this time costs money. The second is the time spent waiting, while native plants grow and stabilise each weeded area in turn. This time costs patience, but no money at all.

Our methods are much more economical than they might appear at first sight: most of our time spent is simply waiting time. The efficient use of light tools reduces both working time and waiting time, since our methods give native plants the best possible opportunity to regenerate rapidly. Careless work at first weeding will cost you dearly. The few hours you may save will multiply into many spent in repeated follow-up; and waiting time will drag on from months into years.

Bringing back the bush is a gentle art, demanding a strong will and patience.

Bringing back the bush.

THE THREE PRINCIPLES

Our method of bush regeneration works on three general principles and you must not deviate from any of them. We cannot stress this enough. Over the years I have found that if you adhere to them in all respects, regeneration will follow. Cutting comers will invariably lead to future problems.

The three basic principles will determine your work plan. Briefly, they are:

1. Work outwards from good bush areas towards areas of weed
2. Make minimal disturbance to the environment
3. Do not overclear.

Principle 1 ALWAYS WORK FROM AREAS WITH NATIVE PLANTS TOWARDS WEED-INFESTED AREAS.

Native plants, if given a fair chance, can and will take back the ground that exotic plants have taken from them. This is the basis of this principle. Natives cannot do this if you work from the weed areas towards good ones.

Where exotics meet natives, there are opposing pressures and the natives give way to the stronger plants from alien environments. However, if the area is left undisturbed, the natives yield ground only very slowly.

To start by clearing the worst areas is not only inefficient, it is actively harmful. Weeds are given ideal conditions when the soil is disturbed and suddenly exposed to full daylight. The balance is tipped in favour of the weeds, which flourish and often spread further into the bush. Such areas are harder to regenerate than those left undisturbed. Weeding a little at a time from the bush towards the weeds takes the pressure off the natives under favourable conditions. Native seeds and spores are ready in the ground, and the natural environment favours the plants that have evolved in it. The balance is tipped towards regeneration. Keep it that way, by always working where the strongest area of bush meets the weakest weeds.

Principle 2 MAKE MINIMAL DISTURBANCE.

This principle applies to everything, both above the ground and below the ground. Work with extreme care. Be careful of the plant canopy, of the roots and of the soil itself.

Vast numbers of weed seeds rain down on the bush every season. If a minute fraction of them managed to grow in undisturbed natural soil, Sydney's bush would long ago have disappeared. Yet wherever it has been left to itself, it is still with us. You can almost always find weeds where developed land meets bushland. How heavily infested the bush is depends on how severely and how often it has been disturbed. In an old suburb like Mosman, where I live, the native undergrowth along the boundaries of the bush has long ago been replaced by walls of weeds. But behind these walls, most wonderfully, the bush has survived. Yes, there are weeds in most parts of it, but not so many that they cannot be cleaned out in one or two operations. Most of the time spent is taken up in finding them.

The areas at the edge of the bush are readily invaded by weeds. The heart of the bush is not. The secret is the seed bed. Undisturbed bush soil under its natural mulch is superbly resistant to weed invasion.

Weeding, however carefully done, leaves gaps and disturbs some soil. It will be some time before the ground settles down into its natural layers: subsoil underneath with its minerals, humus-filled topsoil, then decaying mulch with fresh-fallen plant material forming the topmost layer.

Wherever you work, return soil you disturb as nearly as you can to its original layer, and never forget the mulch. The mulch is your first line of defence against fresh weed invasion. Cherish it, and add to it (see pp. 41–2). When you finish removing a weed, restore the ground – the seedbed of your native plant nursery – as nearly as you can to its natural condition.

Principle 3 LET NATIVE PLANT REGENERATION DICTATE RATE OF WEED REMOVAL.

The better the condition of the bush, the greater the area that can profitably be weeded at any one time and place. Regeneration slows down as the weeds get thicker and the weeding rate must be reduced to match.

Never, never, NEVER overclear. I repeat, bringing back the bush involves two quite different kinds of time: working time and waiting time.

Underclearing has never done any harm, but overclearing invariably costs dearly in both kinds of time. At best, it involves you in a great deal of tiresome and unnecessary follow-up, and at worst it makes regeneration almost impossible. It shatters morale – nobody likes weeding the same piece of ground over and over again with precious little to show for it. If you have a team of workers, people should spread out to weed small amounts in many places. The total area they weed will be just as large as if you had concentrated them in one place, and the regeneration rate will be very much greater.

"Overclearing!? ... No ..."

PLAN OF WORK

At the beginning the effects of working from good areas to bad will go essentially unremarked. Few people notice a scattering of weeds among the bush plants and even fewer notice their absence when you have removed them. On the other hand a large clump of a well-known weed in a prominent position is an eyesore and you will be under pressure to get rid of it. But there must be no compromise. Resist all temptation to abandon the plan of work for the sake of window-dressing. Remove weeds systematically from the bush outwards as described on p. 57. Explain to enquirers what you are doing and why you are doing it. You will find, as we do, that most people will readily understand.

Feel your way from native plants towards weeds and be guided by the strength of the native growth. Do not feel that you must necessarily complete an area at first weeding. Native regrowth helps keep down weed growth and in this way can do a lot of your work for you. If you are at all uncertain about further clearing, STOP. There is bound to be another place where you can profitably take out more weeds and come back to your first area later.

1. Prevent deterioration of good areas

VEGETATION: Native bush with scattered weeds

RISK OF OVERCLEARING: Nil

In good areas of bush, weeds are scattered singly or in groups of four or five through otherwise clean bush. Start by getting rid of these weeds. Practically no follow-up work will be needed. Check once or twice a year, at least once in the flowering season, when colour will show up weeds which you may miss at other times.

Even in a large park, where you cannot hope to cover all the ground immediately, it is amazing what a difference you can make if you pull up one or two weeds every time you go for a walk there. With minimum effort, you can make many hectares of good bush vigorous and safe from invasion for years to come. Regeneration will bring it up against the thicker weeds, which are its next challenge.

WORK FROM GOOD AREAS TO BAD

① Clean native bush

② Native bush with scattered weeds

③ Heavy weeds, some native undergrowth

④ Native undergrowth replaced by weeds

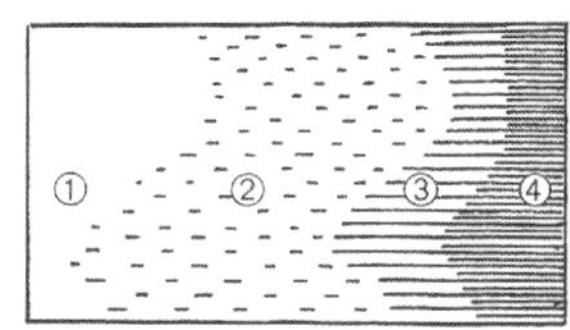

Scattered weeds removed
First pilot strip cleared along the boundary of the heavy weeds (arrow)

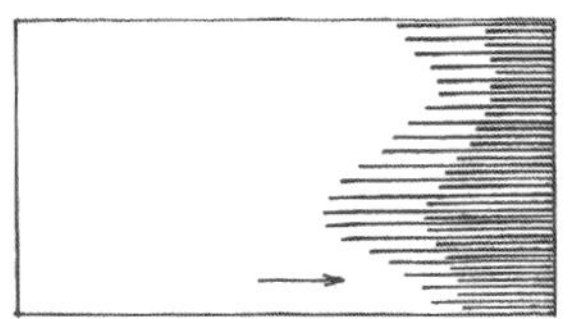

Pilot strip lengthened, in widths varying according to the rate of regeneration

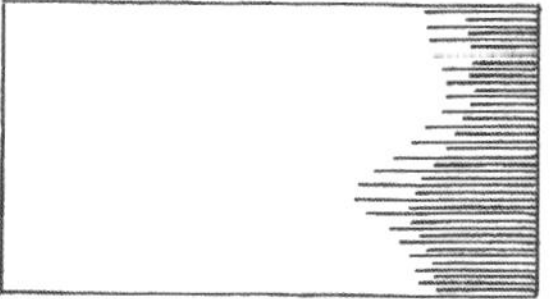

Heavy weeds all cleared, except in one area which has regenerated slowly (arrow)

Natives elsewhere growing strongly

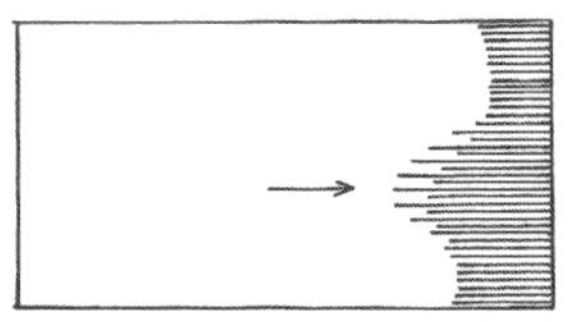

The last of the heavy weeds cleared
A narrow strip cleared where weeds had replaced all native undergrowth
Clearing delayed in slow regeneration area. Clearing deeper in fast regeneration area (arrow)

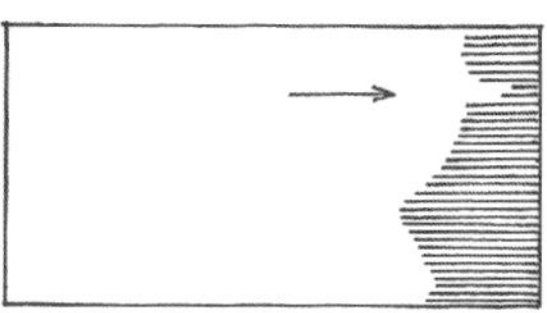

More of the worst weeds cleared, and an 'island' left in the slow regeneration area.

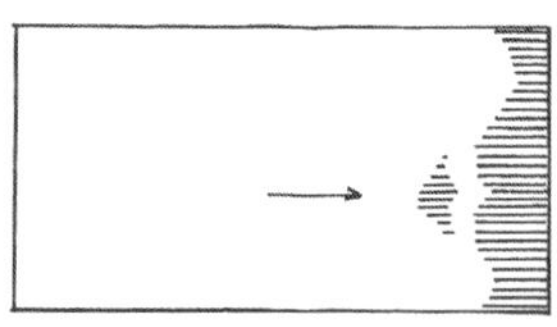

2. Improve the next best areas

VEGETATION: Heavy weeds, some native undergrowth

RISK OF OVERCLEARING: Moderate to high

Once you are confident that you have saved your better bush from deteriorating, you can safely start work on some of the thicker weed patches. Choose a place that you can visit easily and often, where thick native growth is pushing up against a mixture of weeds and natives, preferably not more than one weed to two natives, whether they are seedlings or trees.

I suggest that you start with a strip about three metres wide, and no longer than you can weed about once a month during the growing season. It is unlikely to need as much follow-up work as this, but you cannot tell till you try, and anyway it is quite a thrill to go and see how fast the natives grow when you give them room. If the clean bush/heavy weed boundary runs up and down a steep slope which may erode, work from native shrubs and clear a number of patches, never moving more than about three metres away from vigorously growing natives. This may look a little odd, but regeneration varies from place to place, and however you begin, your strip is not going to stay neat and even.

Let a few months go by before you lengthen the strip. You must expect weed seedlings to appear, as well as natives, and the first weeding after clearing is the worst. It is better not to force yourself into the position of having to do it all at once. If freshly cleared areas are kept small, no great harm is done if something happens to stop you from tending them.

Extend the strip along the boundary, making it longer or shorter, wider or narrower, as experience directs.

3. Hold the advantage gained

VEGETATION: Heavy weeds cleared, new growth a mixture of weeds and natives

RISK OF OVERCLEARING: High

Resist the temptation to start clearing deeper into the weeds before regenerating natives have stabilised each cleared area in turn. The natives need not be very tall, but they usually need to form an almost complete ground cover. Weeds will nearly always keep germinating until this is achieved. The only exceptions we have found were under very dense tree cover. It is the newly disturbed areas that need attention. They will revert to weeds if you neglect them and let in too much light by clearing more ground beside them.

We have found this question of light very important on the Hawkesbury sandstone. It cannot be just coincidence that weed seedlings appear so consistently at the edges of tracks and clearings, even when these are not recently disturbed and are surrounded by hectares of clean bush.

We are able to get away with working larger areas on the Wianamatta shale which, unlike the sandstone, carry a natural herbaceous ground cover. Native grasses and a variety of other soft-stemmed species recolonise very fast and are highly resistant to renewed weed invasion.

4. Cautiously move into really bad areas

VEGETATION: Thicker and thicker weeds, finally reaching complete replacement of native undergrowth

RISK OF OVERCLEARING: High to extreme

When the new growth coming up consists almost entirely of native plants, with only a few weeds among them, it is safe to move deeper into the weeds. Do not feel compelled to do this if it happens to be inconvenient. The beauty of the system is that you can leave a really vigorous, well-established area to look after itself for months at a time.

Keep working along the edge being regenerated, making your new clearings smaller as the weeds get thicker, until regeneration reaches the edge of the worst weeds.

5. Cautiously move into the worst areas

VEGETATION: Very heavy infestation of weeds

RISK OF OVERCLEARING: Extreme

However bad it looks, do not start clearing a block of solid weeds until you have brought good bush right up to it. You may have to resist a very real temptation to overclear. But however much you are tempted, you must wait until the bush has regenerated up to the block of weeds.

The worst weed-infested areas are almost always located along boundaries, paths and at entrances to areas of bush. When faced by a typical wall of weeds on a boundary, resisting the temptation to overclear is not very hard. You have been looking at the weed area for years, telling yourself not to worry about it, so you have had plenty of time to adjust your mind to taking things slowly. But when all that is left to be weeded is one bad patch, it is much harder to be patient. You know that you can clear the patch quite quickly and easily, and complete the primary clearing of a whole big area. It is so easy then to succumb to temptation and overclear, weeding a larger area than native plant regrowth can colonise. I have seen this happen and, four years later, professionals were still paying the price in seemingly endless follow-up.

This does not mean that you cannot start work on a bad area until you have got rid of all the weeds right round it. We start nibbling away at it whenever and wherever we stabilise an area beside it. We poke in around its edges, making clearings as little as two metres in diameter, or we do 'spot regeneration' (see p. 30). We form peninsulas of weeds – even islands – to avoid creating weed-prone open spaces. Be particularly careful not to overclear the strips of bad weeds on the edges of roads and clearings (see p. 29).

At first, after an area is cleared, more exotics than natives germinate and we have to pull up a lot of weed seedlings. In the very worst places where there is not even reasonable tree cover, it may take a whole season for the natives to become dominant, but once they do, the effect is amazing. Just as we are beginning to wonder if the weeds are inexhaustible, the natives take over, and can become very thick on the ground. We have so often seen the disastrous effects of overclearing in similar areas, that even now, used as we are to our own results, we can hardly believe our eyes when we look at our little patches of fresh young natives, growing confidently, hard up against a solid wall of weeds.

In some bush reserves, there are sections cut off from the bush and dealt with too often by mattocks or poisons, where regeneration is almost impossible. Here planting local native trees, shrubs and grasses may speed the process and, if the ground is cleared using our tools and techniques, there may even be some native seedlings germinating naturally. Sowing mixed native seed should also help. But I must admit I can never get as excited by seeing something sown or planted establish itself as by seeing the plants establish themselves naturally in response to our controlled weeding programme. This is true natural regeneration in all its beautiful variety.

It is possible to regenerate very bad areas without planting by bringing the good bush up to them. One such area, the size of a small house block was bulldozed to remove a large chicken run. The coops as well as the topsoil were taken away. By weeding the bush on the perimeter, native growth was stimulated and strengthened. The central area was checked for weeds; some blackberry which survived the bulldozer was removed. It was then mulched heavily in the centre tapering towards the edges to enable the ground covers to spread inwards. Relatively little maintenance was required and within two years the whole area was covered in healthy weed-free bush requiring infrequent maintenance.

Had the bush on the perimeter, however, not been moderately good, the spectacular transformation would not have been achieved in so short a time. This was indeed a case of helping the bush to help itself.

Special cases

Permanent clearings

A permanent open space in or adjoining bushland means an environment permanently favouring weeds. Roads, car parks, playing fields, private gardens, even small clearings made for park seats, are all potential trouble makers. If a strip of soil is disturbed and left exposed and particularly if filling has been brought in to level the ground, weeds can quickly replace the natives.

Tracks spread out from clearings, particularly from places where people leave cars. Many people, who would not dream of walking over a garden bed, will tramp unheedingly through the bush, smashing the plants and compacting the soil. Weeds are spread further, let in by the destruction of the native undergrowth. All our worst infestations have spread out from clearings, often accelerated by rubbish-dumping.

The verges of clearings are difficult places. Even when a strip of weeds is only a metre or two wide, clearing it in one operation seldom brings the bush outwards towards the clearing. In fact, very often it lets the weeds, which are the more vigorous growers, in further. When weeding in such places, follow principle 1 and work as always from the bush towards the weeds. You do not have to weed in a continuous strip. Weed instead a number of separate wedge-shaped areas, with the thin ends of the wedges pointing towards the open space. In this way you will break through the screen of weeds at scattered points, making small openings which will not let in much light. Make the wedges wider as the bush regenerates and they will join up to form clean bush (see p. 25).

It is important, particularly when working where the weeds are fairly thick, to remember not to open up spaces in such a way that people will walk through the gaps you have formed. It is frustrating to have to leave a screen of weeds until the natives grow thick and tall enough but, if the way looks clear, and people use it as a path, sparse small natives will be destroyed by the tread of feet. Remember, too, that if you yourself leave even a faint track, others will probably follow it. You can use the larger weeds, laid down with their roots in the air, to form a most useful barrier and prevent the public from trampling the ground. We have closed a number of trample tracks by blocking *both ends* with uprooted lantana.

Spot regeneration

Eileen worked often on a dry headland which has a lookout with park seats near its top, is handy to a car-park, overlooks a popular fishing spot and is bounded on one side by a busy walking track. Some of the worst weeds are cut off from clean bush by tracks and rock outcrops. The area slopes very steeply to the water. Trampling is a major problem and people scrambling up and down it have even caused a small landslide. It is also infested with rabbits. A real problem area, yet now it is regenerating.

In places like this, clearing any sort of strip, however irregular, invites disaster, so my sister evolved the special technique of spot regeneration. When she found a native plant surviving among otherwise almost unbroken weeds, she gave it *room to grow,* often at first by removing just one large weed plant. Released from competition, the native grew much faster than before and later other native seedlings appeared. This made it profitable to remove a few more weeds and enable the bush to extend further.

In this way she encouraged patches of native plants to grow throughout the weeds, never laying bare a patch of soil big enough to erode and never clearing an open space big enough to invite trampling. The rabbits devoured everything green, weed and native alike, so native growth was slower and less varied than usual, but heavy weed growth gradually became lighter and, suddenly a season or two later, the patches of natives coalesced to form clean bush. The transformation of the headland was visible from the other side of the bay.*

Creek banks

Creek banks must be treated with extreme care. It is dangerous enough to work beside a permanent open space – in this case the creek itself – without having the work made doubly risky by the two powers of running water: its power to attract people and its power to erode soil. People will take advantage of any opening that allows them the merest glimpse of a stream, and floodwaters will wash away any soil that is not bound fast in place by the roots of growing plants.

These troubles can be avoided by combining the two strategems described above. On the upper part of the bank, make very small wedge-shaped openings, apex towards the creek; and from top flood level to the creek bed, carry out very cautious spot regeneration. Then possess your soul in patience, and wait. If the bank is very weedy, progress will be slow but it will be very sure, and you will find great pleasure in watching small natives grow bigger, and new ones appear, spreading their roots and holding the banks against both weeds and water.

* *Editorial Note:* Whilst Joan lists spot regeneration as a special case it is actually an excellent example of the application of the Bradley method in microcosm.

FIRE

Fire is a major disturbance to the bush. It causes drastic changes, for better or for worse. Either way, burnt bush, by demanding attention soon after the first good rain falls on the burnt ground, thoroughly disrupts your nice methodical plan of work. How much disruption and for how long depends on whether the bush has burnt fast when the ground was hot and dry, or slowly and steamily when the ground was cool and moist. The effects of the different types of fire are very marked. Generally, with quick hot fire the seeds burst open, still allowing germination to take place, whereas in a slow burn, pods become too limp to open and the seeds inside are stewed and so lost. Who knows what damage the steam may have done below the ground level?

Areas affected by wildfire

A blazing hot wildfire is a terrifying thing, and its effect on the bush seems devastating. Trees stand leafless, their blackened trunks stark against the grey ash which is all that remains of the smaller plants.

The media regularly report that the bush has been destroyed. It has not. Our native plants are very well adapted to survive *occasional* wildfires, needing nothing but water to make immediate use of the mineral wealth hidden in that dead-looking ashbed. They respond to the first good rain with a joyous surge of growth from roots and seeds, trunks and branches. Wattle and pea seedlings abound, their root bacteria gathering from the air in the soil spaces precious nitrogen for themselves, their hosts and all the other plants as well. Sooner than seems possible, the bush is fresh and green again.

Even the slow-maturing species are not permanently lost but, given time, come to full flower and set ample seed. Some species, such as boronia, can take five years to recover fully; others, like stenocarpus, much longer. However, if further fire intervenes before the plant has had time to mature and set seed, the whole species can be wiped out.

One of the many myths about Australian vegetation is that occasional fire is essential for its renewal and that only natives are affected in this way by fire. But natives are not the only plants to flourish on an ashbed. I do not know of a single bush-invading weed which does not respond in exactly the same way. Seeds germinate in abundance, and garden plants, like dahlias, lasiandra, cotoneaster, lantana and privet shoot from their base like gum trees.

Even so, wildfire can be turned to good effect if you get in early and tackle the weeds whilst they are still small. You will be blackened by charcoal but do not lose heart. Regeneration of native species will be spectacular and at the end of twelve months, weed germination will be almost zero.

Prescribed slow-bum areas

Hazard-reduction fire is completely different from a real bushfire. It is sometimes called a 'control' burn, a term which has always seemed to me rather a misnomer in view of the number of times such fires have got out of control.

Burning is done at the time of the year when the bush is dormant, and on the ground the leaf mulch is heavy and moist. For this reason it must be deliberately lit, frequently many times, whereupon it smokes, smoulders and steams. This type of burn consumes only the understorey, leaving most trees and many of the taller shrubs still green at the top. The ground smells like a garden rubbish heap, not a bit like bush after a bushfire, because the mulch burns only on the surface, while the underlayer of mulch is left to steam at high temperatures, and then putrefy.

It should be remembered that in the cooler seasons, plants and trees are storing food in their roots, ready for the surge of growth in spring, so this is not just an unnatural and unwelcome interruption to the vital feeding process. It is a major setback, as it was intended to be.

Even those dedicated burners, the Western Australian foresters, have had to admit that their repeated 'gentle' fires, far from encouraging the growth of timber, have contributed to the dreaded Jarrah dieback. The mycorrhiza and the important nitrogen-fixing legumes, along with the hibernating lizards and smaller creatures caught unawares, have been destroyed. The soil is not ash grey but black. A prescribed burn has a disastrous effect on native plants and an absolutely explosive effect on weeds. With the understorey gone, the soil, now changed for the worse, is exposed to light which weeds thoroughly enjoy. The conditions favour the growth of weeds, which keep germinating, and the slow and uneven growth of the native plants does little over the years to keep them in check.

In 1965, a copybook slow-burn was carried out in Ashton Park. The area was, for Mosman, a very good one, with only scattered weeds except for one patch of heavy lantana that did not burn at all well. The area could almost all have been stabilised before the fire by one weeding with minimal follow-up.

Cleaning up after the burn fell to the lot of my sister and me. There was nobody else to do it. We began to think that we would never be able to relax, as different weed species followed one another in a seemingly endless succession and sometimes returned years later. It was rather like painting the Harbour Bridge as we systematically covered the 2½ steep hectares, from bottom to top, from end to end, in 10 metre strips and then went back to the beginning to start all over again.

This went on for eight years. The weeds were fewer after the first three, but still relentlessly appeared over the whole area. We got extremely sick of it, especially as we kept getting tangled up in vines; shrub regrowth was so poor that they had nothing to climb on and were draped about on the ground. The natural balance had been totally upset, not just among the flora, but also among the fauna. My sister repeatedly asked, 'Where are all the ants? Why aren't there any ants here?'

Slow-burns mean a long, slow weeding, which can be most depressing but which must not be neglected. If you have to attend to burnt ground, the only safe course is not to start any new work. Confine yourself in other areas to any follow-up necessary to prevent your losing ground there. If you have been working obediently and faithfully according to the principles, rules and plan of work, this follow-up will mean no more than a quick check-around and not a great deal of work.

Firebreaks cleared by council labourers invariably become weed-prone areas, but selective hand-clearing of dead wood, some native shrubs and part of the mulch will do far less harm than burning. If council thinks that selective hand-clearing is too expensive, you may have to use volunteers. Such work has been successfully carried out in different parts of Sydney under the direction of an experienced regenerator.

So use all the eloquence and influence at your command to dissuade the powers that be from slow-burning your bush. Remember that even those most dedicated to the slow-bum do not claim that it eliminates fire hazard, but merely that it *reduces* it.

the mosses,

the ground covers and ...

the living mulch so vital to the food chain.

Kikuyu taking over after casuarinas were killed in a slow-burn.

Mulch from casuarinas provides a good weed barrier, especially valuable along road verges.

Ashbed at Rawson Park (now Bradley Bushland Reserve) after wildfire and ...

Native regeneration.

Ashbeds

Burnt ground with ash still lying on it we call an 'ashbed'. These areas must be weeded and kept weeded or the exotics will win the battle, served well by their very nature – the vigour with which they grow.

Ashbeds are most gratifying places on which to work, but there is no denying that, to get the best out of them, you need to do a lot more work in a much shorter time than usual. The hotter the fire has been, the greater will be the number and variety of seedlings that you can expect to find, and the faster they will grow. Weeds, both regrowth and seedlings, are easier to find and remove while the ground is still almost bare, but do not worry if the burn covers too big an area for you to clear all at once. The minerals in the ash affect growth, and the effect can be dramatic and persistent, lasting for years.

Weeds follow 'Slow-burn'.

RULES FOR WORKING IN THE BUSH

1. Watch your feet

Be very, very careful how you move through the bush. Plants do not like to be trodden on, and natives in particular hate compacted soil. Careless trampling does a lot of damage and the effects last many years.

This is the hardest rule of all to teach and novices, concentrating hard on using unfamiliar tools and techniques, do not find it easy to remember that encouraging the natives to re-establish themselves is the object of the whole exercise. They will triumphantly extract a weed, then step back on to a native seedling. Later at lunch they may forgetfully sit down on a native grass. It can take quite a while before it becomes second nature to watch where you put your feet, your free hand, your knees and your behind.

When I say 'Watch your feet', I mean it literally. Never hurry. Watch where you place each foot, making sure that it does no avoidable damage. Remember you are not a soft-footed native animal, but a large and heavy human, whose whole weight on a hard heel bruises plant and soil alike. Wear soft shoes. Put your toes down first and ease the weight on to your heels, treading lightly. Where native plants are small but growing thickly, there does not seem to be anywhere to put down even a careful foot. However, if you insert a toe into the thinnest patch within reach you will usually find you can bring your heel down safely. Have you ever watched herons fishing? Your movements should be much like theirs. They make hardly a ripple on the surface water and do not stir up the mud at all.

Be very careful on steep slopes, particularly when going downhill. If you let your foot slide, it will carry with it leaves and twigs and even soil. Such damage is hard to repair.

A small weeding party moving in single file can open up a trample track in one operation. Others will follow it, curious to know where it leads, and the scar becomes permanent. So people should spread out, taking different routes, and incidentally on the way, finding and removing small weeds that would otherwise have been missed.

Move about no more than necessary. If you are trying to remove a big weed that will take you some time, settle yourself – standing, sitting, squatting or kneeling – in a comfortable position and stay there until you have dealt with everything you can reach. If you want to ease cramped muscles, straighten up but do not 'mark time' by lifting your feet.

Look behind you before you step back. Remember that any native you tread on will suffer a setback and may even be killed. Never lose sight of the object of the whole exercise – to encourage the growth of native plants.

2. Disturb the soil as little as possible

We prefer hand-pulling weeds to using any tools at all. If we have to use tools, we never use a heavy one if we can use a light one. Many widespread weeds, such as lantana, boneseed and honeysuckle, are surface-rooted and surprisingly easy to pull up, even when they have grown beyond the seedling stage. If cut root-ends are left well buried, they do not grow again, so if a weed is too big for hand-pulling, we remove it by tracing its roots until we can safely cut them below the soil surface, using the smallest tool that will do the job.

A sheath-knife, secateurs (each sharp), trowel and pliers will account for all but the very largest weeds. On these we use a pair of loppers, a hatchet or, in extreme cases, an axe – the last near the main stem only, where roots are at their greatest thickness and native plants are hardly ever present. Vertically-growing roots need to be cut at a much deeper level than horizontal roots. For these, we use trowels and our fingers to dig deep but narrow holes. Here pliers often give a better grip on a root than fingers. The very deep main roots of ochna and camphor laurel will usually regrow even if cut and buried deeply, and must be totally removed.

When you are not used to it, following out roots one at a time can be a tedious process and you may long for some quick action. There is a splendid air of speed and efficiency about machinery and heavy tools but they damage more than weeds. They cut the roots of nearby natives, kill small native seedlings and destroy the natural structure of the soil. In the long run they cost more than our light tools, in both working time and waiting time. It simply does not pay to use them.

3. Preserve and replace the mulch

Call it mulch, not ground litter. The unpleasant connotations of the word 'litter' does not suit the lovely stuff that is so necessary to the bush you are regenerating. It is chock full of living things, some of which are essential to the healthy growth of native plants. It holds moisture, prevents erosion, and protects the soil from sudden changes in temperature. As it decays, it returns plant foods to the soil. The native plants love it and incredibly few weed seedlings come up through it. It may look a muddle, but it really is not. It is a highly organised series of interlocking layers, teeming with the innumerable organisms vital to the natural cycle of birth, death, decay and rebirth.

Look at its structure: it has not come about just by chance. On top it is a loose, springy, crackly mixture of flowers and fruits, leaves and twigs, bark and branches, with here and there a fallen log. Lower down, you can see the beginnings of decay. Petals are unrecognisable. Thin leaves are thinner and are starting to break up; some show their beautiful skeletons. The colour of the mulch is changing from light to dark brown. The mixture is moister, more tightly packed. Where drifts of leaves have collected round a fallen log, its bark is a maze of insect tunnels. Lower still, merging into the earth and forming part of it, is the soft dark wholesome-smelling blend of plant and animal material which we call humus. Unpaid workers have been busy, transforming dead matter into living soil. Predators and parasites are important too. There they all are, feeding on the mulch and on each other, dropping their excreta, committing their bodies to the soil. Wastes there are, but nothing is wasted. Beetles and earwigs, ants and termites, springtails and cockroaches and spiders – breakfasts for birds and bandicoots – things like these you can see hurrying for cover when you disturb their shelter.

But you cannot see the millions and millions of smaller organisms, right down to simple protozoa and bacteria which are essential to the whole process of transforming mulch into humus. You can see bracket fungi, toadstools and their white threads of feeding mycelia, but not the slender filaments of the mycorrhizal fungi that feed plants by entering their finest root hairs.

Some organisms like plenty of air and a little light. They live in the top layer. Others like less air, less light, more moisture, a more even temperature. They live lower down. Others may live at different depths at different stages of their life cycles, or move continually from one layer to another, very gently stirring the mixture.

However carefully you weed, you cannot avoid damaging this complicated fabric. Never forget to repair the damage as you go, smoothing and lightly spreading the mulch back to form an even cover. Then the living organisms in it can quickly go to work again.

4. Mulch with the weeds themselves

Burning weeds or carting them out of the bush is worse than useless – it is wasteful. We keep everything we possibly can, to add to the mulch. In dry places we just put most pulled weeds on the ground with their roots in the air; in damp places we hang them up to dry on the nearest native tree or shrub.

There are a few weeds, or parts of weeds, that are simply not safe to leave in the bush. These we do carry away, but they are only a very small part of the mass of vegetation which we put in safe places until it is ready to atone for its misdeeds and add itself to the protective mulch and give back to the soil all that it has taken out of it.

5. Do not pile weeds in heaps

Piling weeds into neat, easy-to-carry heaps makes sense in a conventional

Some weeds must be carried out of the bush.

garden, but it is bad practice in the bush. Heaps of soft weeds rot down into a nasty squashy mess that is quite the wrong environment for natives, and there is a very good chance that some weeds will re-root and flourish. Heaps of woody weeds are an awful pest to untangle when, as they often do, some of their seedlings grow up through the heaps.

So disperse what you uproot. The soft weeds will quickly dry out and the woody ones will not get in your way during follow-up.

6. Never hang weeds on weeds

Hanging weeds on other weeds puts you to the trouble of hanging them twice. I have a vivid recollection of working on a heavy weed front, where a thoughtless weeder had thrown lantana canes forward on top of other lantana. The living canes had grown up through the dead, forming a thorough entanglement. It was the work of the world to get it all out of the way to reach the next strip of roots. This was an extreme case, but even one weed hung on one other means double handling and that means inefficiency. So hang your weeds on natives, not on their fellow-weeds. You will often leave recently treated sites looking terribly untidy, but do not let it worry you. Look down. The soil surface itself will be absolutely impeccable, covered evenly with the mulch you have carefully preserved and replaced, ready to welcome the regenerating natives which are the object of the whole exercise. Above ground, leaves quickly wilt and their stems start to droop. This is a splendid time to show off your work to visitors – it all looks so much more convincing than it will a year or so later, when all you can do is to tell them that the clean bush they are looking at was once a mass of weeds.

7. Remove all species of exotics from areas weeded

Never single out a particular weed species as your pet hate, forgetting all others in your pursuit of it. At present the newsworthy plant is privet; a few years ago it was lantana. Site after site is consequently dominated by weeds like honeysuckle and wandering jew, which are much harder to control than the more obvious weeds they have replaced. It is a waste of time to kill one weed to make room for another that may quite probably be worse. I learnt this lesson very early when I found a fine collection of young privets about half a metre high, growing with a few harmless looking wisps of wandering jew among native ferns. The privets were certainly doing the ferns no good, and I thought that it would be easier to get at the wandering jew if I got the privets out of the way.

They were easily pulled up and I went off happily, intending to return in a day or two to finish the job. A succession of heavy rains stopped me and it was weeks before I got back to the steep and slippery gully slope on which I was working.

The ferns were mildly grateful to be relieved of the privet pressure, but the wandering jew was absolutely delighted. It was all over the ground, and had long and very brittle tendrils well tangled up in the ferns. My few minutes indiscretion cost me hours in follow-up. So remove all weed species and leave the ground completely free for natives.

8. Work with the weather

Like it or not, our work is entirely at the mercy of the weather. Is it ever going to rain? Is it ever going to stop? If the weather is too dry, fine soil sets hard, and coarse soil flies about when worked. Neither will pack back properly round the roots of natives. Even if it made sense to take twice as long as it should to get out a few weeds, there is a risk of erosion when drought-breaking rains arrive at last. If the weather is too wet, every step you take means soil compaction, which inhibits native growth. Sloppy lumps of soil come up, clinging hard to weed roots, resisting replacement. Either way, work is at best unprofitable, at worst actively harmful.

Even in small reserves, soil types can vary widely and you will soon learn which hold moisture and which drain quickly. But eventually you will find yourself forced to give way to drought or flood with the best grace you can muster.

9. Do not remove any plant you cannot identify

It is not good enough to *think* that a plant is a weed. Even if you are right, it is not going to overrun the bush in the time it will take to have it properly identified; but a pulled-up native is lost for ever. Get to know your plants, as explained below, and always remember that it is better to be sure than sorry.

GETTING TO KNOW YOUR PLANTS

It is essential to make sure that you do not pull up any natives, or preserve any weeds. *A weed is a plant out of place.* This traditional definition cannot be bettered. All exotics, whether from overseas or from other parts of Australia, are out of place in our natural reserves.

We are working to bring back natural bush, which to our minds needs no 'improvement', and we regard all local natives, whether pretty or not, as good. They are in their own environment, where their ancestors have been getting along together since the last ice age, and time and natural selection have kept them in balance. So we make no attempt to control native parasites or prevent rampant native creepers from bringing down shrubs or old trees. They are all part of the ecosystem and will give way to other plants when their time comes.

Some weeds look very natural in the bush and some natives look so wrong that it seems hardly fair. Since it can be easy for amateurs like us to misidentify plants, we collect a specimen of every species we find in the bush and have it checked by the National Herbarium in the Royal Botanic Gardens, Sydney. If we can, we identify it ourselves by using one or all of the increasing number of books available on the subject. There is nothing like wrestling with a formal botanical description to make you look really hard at a plant. Often, we are sure we have identified it correctly, but sometimes we are not sure, so, even though we are now very familiar with most plants, we still do not skimp the checking.

One or two people working alone can coast along excellently, learning their plants as they go, as my sister and I did. We did not need to memorise the names of hundreds of different natives all at once. All we needed was to be sure of the weeds. Most of them do not invade the bush, but are found in badly disturbed and neglected places, where people dump garden rubbish. Away from such places you will mostly be pulling up only a dozen or so different species and I can assure you that in time you will recognise them, even when they have no more than two leaves.

For professional work, each weeding team must include at least one member who has a good working knowledge of the local plants, both exotic and native. Knowing the weeds is essential; knowing the natives is very helpful indeed.

Joan with her lens at Parrawi Park. COURTESY OF THE BULLETIN, 6 FEBRUARY 1980.

If you can accurately identify most of the local natives, at least you know they are not weeds! Also, you will be repeatedly asked to name them when visitors come to see what you are doing, and you will lose face if you are unable to do so. This is not to say that we cannot still quite often be stumped, particularly by grasses, which baffle most amateurs and many trained botanists as well. Seedlings, too, are often difficult to identify until they develop leaves more typical of the mature plants, or until they produce flowers or even fruits.

The great thing is to realise that you do not know everything. If you are in any doubt at all about a plant, leave it in the ground until you have had it properly checked by a competent botanist.

In New South Wales, and probably in other States as well, it is illegal to remove any kind of plant material from a public reserve. Indeed, you could be prosecuted for merely collecting a specimen of a declared noxious weed, let alone digging it out. So if you are working as a volunteer, you will need official permission, not only to work in a reserve, but also to take specimens for identification.

It is extremely hard to recognise a plant that has simply been allowed to shrivel up and die, so if you want to stay friendly with your experts, give them specimens that are either completely fresh or properly dried. Drying is by far the best. It yields you an invaluable collection which you will find you continually refer to, particularly for identifying plants you seldom encounter and difficult groups like the grasses. Furthermore, dried specimens can be sent away by mail for identification. Indeed, our mother, when away on visits, used to send flowers to us folded up in her letters home, and both arrived in perfect condition.

PRESSING AND MOUNTING SPECIMENS

To make a plant specimen collection to enable you to identify plants, you need to be able to dry, mount and store specimens.

Drying the specimens

You can dry most plants by simply putting them between the leaves of any reasonably solid book. An old telephone book is excellent for this purpose, but it is more convenient to make your own plant press – a simple and highly effective device which will cost you practically nothing. All you need is plenty of newspaper, two reasonably rigid outer covers and straps to hold everything together. Two sheets of hardboard make excellent covers and exert mild pressure when held together with the straps. The equipment does not crush the moisture out of the plant tissue. It transfers it gently to the newspaper.

Specimens for pressing must be fresh and unwilted, in which you can see the shape of both leaves and flowers. Arrange each specimen to show its essential character: the parts of the flower and both the underside and the upperside of at least one leaf should be well displayed. Tall grasses, of which you need to preserve a complete stem, should be bent into V- or M-shapes to fit them onto the paper.

Setting up your press

Fold a sheet of newspaper. This becomes the folder for your first specimen. Write on it the name of the specimen, where and when you collected it and add a few brief notes, such as growth habit, type of bark etc. Put several sheets of folded newspaper on top of one of the hard covers. These folded sheets form a dryer and absorb the moisture from your specimen. Now put the folder with your specimen in it on top of the dryer, add another dryer, and then another specimen in its folder, and continue alternating these layers until you have all your specimens ready for the press, finishing with a dryer. Put the top cover on and strap the whole press together.

Most native plants will dry out in a week or so, but many exotics and a few natives contain a lot of moisture. You will have to change the dryers around such specimens night and morning for the first few days, to prevent the plants from going mouldy.

Mounting the specimens

When fully dried out, woody plants become stiff, and soft herbs are paper-thin. You can use school drawing paper to mount specimens on, taping them in place with strips of glued paper about 3 mm wide. Do not use ordinary stickytape to hold the specimens in place. This dries out and will eventually fall off.

Filing the specimens

You can use ringbinders to hold the mounted specimens, putting each one in a plastic sheath for safekeeping. If you wish, you can take a whole binder out into the field with you for reference. In this way you can avoid collecting specimens unnecessarily.

Themida australis (kangaroo grass). Pressed for mounting.

EQUIPMENT

I carry seven items: knife, trowel, secateurs, pliers, a bag for weeds, gloves and log book. I seldom work a three-hour session without using all of them. These are the basic tools which you will use to remove the small and medium-sized weeds. Each regenerator either owns a set of tools or is issued with one and becomes responsible for bringing it to work and keeping it in good order. Carry them with you at all times in sheaths, or in a pouch stitched vertically to make separate compartments, or in pockets (always remembering to have your knife well sheathed for safety's sake). I personally prefer to keep them separately, partly to spread the weight and partly to be able to put a hand easily on the one I want. You can make your own choice but you do need to have some container for carrying tools in which will allow you to keep your hands free.

1. Knives

Of all the tools, the ones we use most are the knives. We cut with them, dig with them, scrape away soil with them, rearrange ground cover and mulch with them. Given time, a suitable sharp knife will deal with all but the very biggest weeds.

Any tool well suited to its purpose is comfortable to use. Your knife should have a fairly thick handle, which you can grasp without cramping your fingers, and with a broad, smooth end which will not hurt your hand

when you thrust the blade into firm soil. The blade should be of good steel and very robust. Working in soil demands much of a knife. Poor steel will lose its edge in half an hour, and there is real danger of being badly cut by the broken end if a flimsy blade breaks under strain.

Replace your knife when necessary. Between abrasion and frequent sharpening, the best of knives wear down in size to the point of inefficiency. Also, they are easy to lose. I once dropped one in heavy mulch on a steep slope. I actually saw it leave my hand, yet two of us searched for it long and diligently, and quite in vain.

It is not easy to find the right knife at the right price. After a long search through the shops, we have settled on the knife illustrated – a butcher's wide, boning knife.

2. Trowels

It is just as important to have the right kind of trowel as it is to have the right kind of knife – you will be using it very nearly as often. It should be long and narrow, so that you can push it deep into the soil, and it should be only slightly dished, so that you can probe with it among rocks and between the roots of natives. To make it easy to push down beside vertical roots, and to reach into awkward places, the blade should be almost in line with the handle.

Since we have yet to find a brand ready-made to fit all these specifications, we buy the trowel known variously as bulb, fern, transplanting or seedsman's trowel, grind down the tip to the shape illustrated, and straighten the kink where the blade joins the handle.

Choose a brand with a strong blade, of thick metal that fits around the outside of the handle. Blades that are just crimped and pushed in are flimsy and have the nasty habit of falling out of the handle. If you do not own or cannot use an emery wheel for grinding down the tip, your local hardware merchant may be able to grind it for you. Give him a tracing of the tip to make sure that you get the right shape. This is critical. Too blunt a tip will not slide easily into soil or fit into small spaces, but too fine a tip will bend or even break. Smooth the ground edges lightly with a file so that they are not too sharp. To bring blade and handle into line, put the blade in a vice and lean rather carefully on the handle. This operation also serves as a reminder that the sturdiest of trowels is still a light tool, which will bend if you treat it too roughly in the bush.

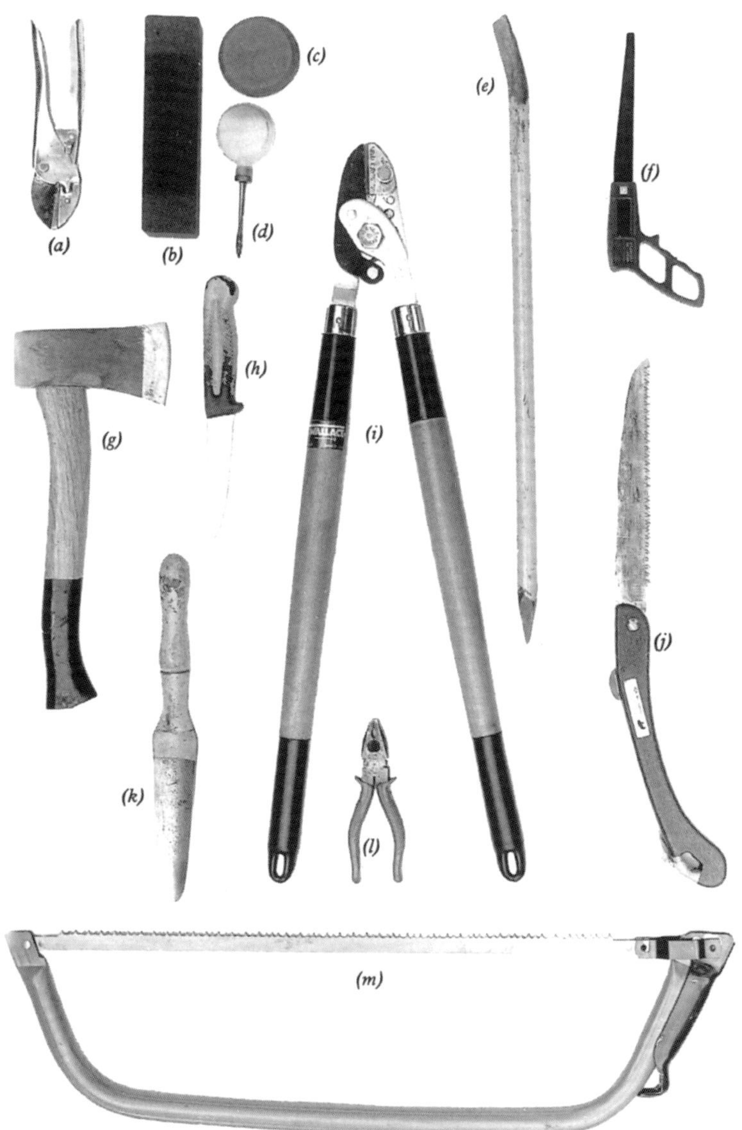

(a) Secateurs *(b)* Oil stone *(c)* Axe stone *(d)* Oil can *(e)* Jemmy *(f)* Miniature saw *(g)* Hatchet *(h)* Knife *(i)* Loppers *(j)* Folding saw *(k)* Trowel *(l)* Pliers *(m)* Bushman's saw.

3. Secateurs

My sister Eileen had a definite aversion to knives and greatly preferred secateurs, which she used with high efficiency. Though not as versatile as knives, they do have certain advantages. They cut more easily through the thicker roots and stems, and can be poked into confined spaces where it is awkward to put the necessary pressure on a knife.

They can also be used for digging, but remember to hold them firmly closed to avoid getting the blade clogged with dirt. The traditional secateurs have curved blades that have to be sharpened by a specialist. So use the straightbladed 'anvil' type secateurs, which you can sharpen yourself. Get a good strong pair with all-metal handles. Plastic handles, in our experience, have a nasty habit of falling to bits. Use plastic keepers to hold the blade closed when not in use. Another aspect to consider when buying is the ease with which the blades can be separated for sharpening. Try to buy a model that does not call for the use of a screwdriver (or even, in one I have seen, *two* Phillips head screwdrivers).

4. Pliers

These are the least used of the four tools. They are invaluable for pulling stubborn roots like paddy's lucerne or for gripping roots where there is little room to manoeuvre or to get a firm hold on them with your fingers. Remember that your fingers are sensitive enough to feel a root begin to give, but pliers are not. Be careful how you use them, or you will find yourself breaking roots instead of teasing them from the soil.

5. Gloves

Bare hands are the only thing for dealing with fragile weeds like wandering jew but, unless you have an unusually tough skin, you will work on most weeds much more comfortably and just as effectively in light well fitting gloves of cotton, leather or plastic. They must be soft enough for you to be able to feel through them for concealed roots. Heavy gloves leave you practically no sense of touch and should be reserved for use on very thorny plants like blackberries.

6. Weed bags

Grocery bags make good weed bags. We are currently recycling our grocery bags, sometimes putting one inside the other to prevent seeds from falling through tears or holes. The weight of weeds in bags of this

Regenerators with the tool boards.

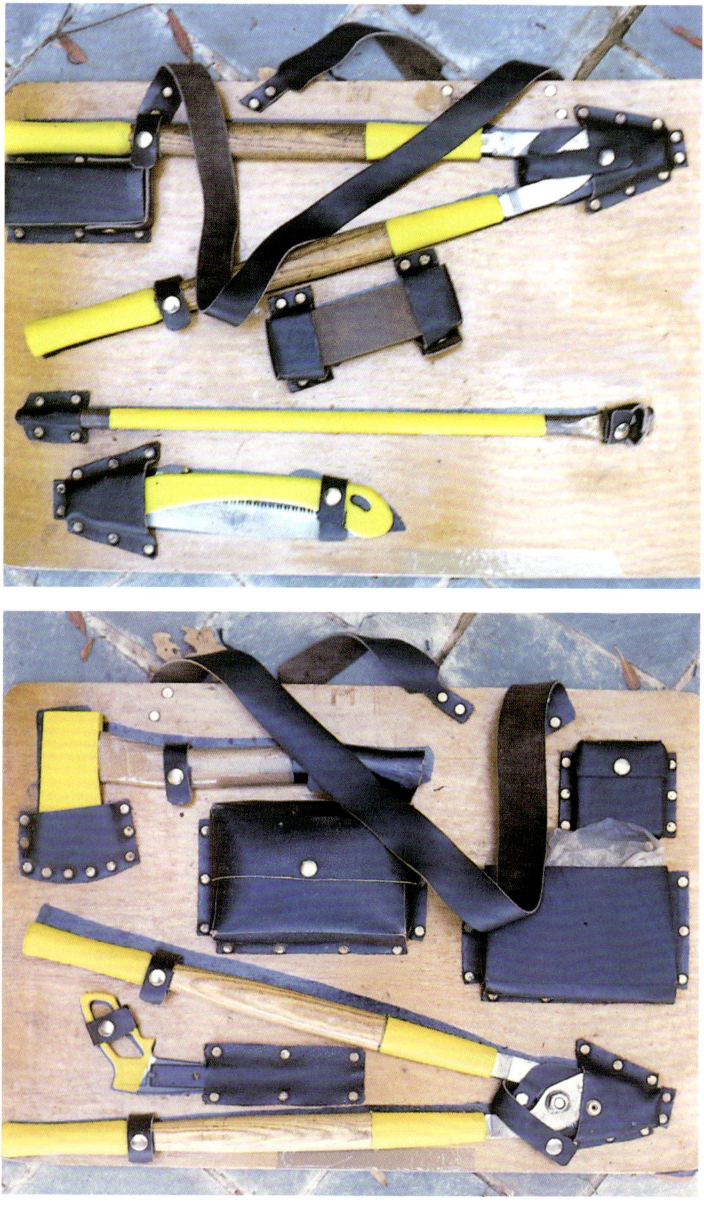

The tool boards.

Removal of the weeds in the foreground can save months of work later.

Erosion caused by spraying. Note absence of ground cover.

size will do less damage than large laden bags, if you accidentally put them on native seedlings.

Succulents, seedheads of annuals, bulbs, tubers and asparagus 'fern' berries (both green and red) cannot be left in the bush. Neither can plants that root from every node, such as lambs' tails or madeira vine, nor wandering jew which breaks at a touch, each bit forming a new plant. Honeysuckle and morning glory, grasses such as kikuyu and buffalo once uprooted can all be rolled up and hung securely on native plants to dry out. Small pieces, however, which cannot be securely hung, must be bagged and carried out of the bush with you.

7. Logbook

Keeping a log book should be the responsibility of one person in each group. It is your document for future reference – map of reserve, site, time spent on weed removal, weed species removed and progress of native plant regeneration. It is an invaluable source of information (e.g. weather, evidence of rubbish dumping, fires, clearing or removal of plants or soil,

Bagging seedheads.

construction of paths, new residents, etc.) as well as interesting and rewarding to look back on and to show to unbelievers.

Members are encouraged to carry a small individual notebook in which they can jot down any information that will help them identify plants they encounter.

8. Toolboards

These are used for transporting the larger tools described below. The toolboards we use are convenient, manoeuvrable and easy to carry. They are just sheets of ordinary three-ply with a carrying strap that has a half-tum in it to let it lie flat on the shoulder. The outline of each tool carried is painted on the board.

I have come to regard toolboards as essential if you are using tools that have been supplied by the management. It is easier to select tools from a board than to have to root about in a box for the one you want, and the painted outlines on the board are an effective way of keeping check that all tools have been replaced before you leave an area.

9. Lopping shears

These powerful tools work on the leverage principle. Unlike brushhooks and mattocks they cut only what you decide to cut, and are powerful enough to deal with anything up to the size of a small tree. They are a nuisance to carry about from weed to weed and are easy to leave behind if you put them down, but their big advantage is that you do not need much room in which to wield them. Two pairs are generally ample for four regenerators.

Although they are not very easy to find in shops, the double-leverage ones illustrated are excellent. Do not get anything bigger or the advantage you gain in extra leverage will be lost because of the difficulty of spreading the handles in a confined space. They have to be taken apart for sharpening and as putting them together again afterwards can be quite confusing, keep the diagrams and instructions that come with newly bought ones.

10. Hatchets and axes

These are used on trees and also on shrubs with roots too thick to be cut with loppers. The hatchet illustrated weighs 600 grams, handle and all. The blade is 140 mm long and 80 mm wide. Used carefully, the narrow blade is selective enough in what it cuts to be able to be used on the roots of privets which are growing among pittosporum species.

We use the hatchet in preference to the axe. With the hatchet you can work sitting or kneeling to cut roots below ground level, which is impossible if you are using an axe, and the short handle of the hatchet is a great advantage in a confined space. The axe can be dangerous to wield and we use it only when cutting large tree roots or those of big, deeply-rooted shrubs.

11. Folding saw

This saw is really a bushman's folding saw with reasonably large teeth. It makes short work of timber too heavy for loppers. For safety and ease of carrying we selected this rather than the more conventional bushman's saw.

12. Miniature saw

On a dry rocky slope in Ku-ring-gai Chase, a group of regenerators found an inordinate number of lantana plants, whose stems were wedged between horizontal slabs of sandstone. It proved almost impossible to get the usual tools past their multiple stems and swollen butts to reach their roots, some of which were very big. One frustrated regenerator suggested a small saw, but another went one better. He had brought a multiple-bladed pocket-knife which included a saw with quite coarse teeth. In a case like this, carrying an extra tool is well worthwhile.

13. Jemmy

The jemmy or bar can be used to raise heavy boulders or ease rocks away when tracing out roots. It works with greater efficiency if the claw is removed and the end tapered to a point.

14. An oilstone and oily rag

These are carried for the sharpening of knives and secateurs, which is constantly necessary.

15. First aid kit

16. Notched spade (the 'agaverer')

This tool has no space outlined on the toolboard. Eileen and I devised it ourselves to deal with a nasty infestation of very large century plants *(Agave americana)*. Their leaves were so hard to cut away we had difficulty getting to the bases of the plants. The densely matted fibrous roots defied the knife and turned aside a well sharpened spade. Pride and principle both forbade our using a mattock. The notched spade, christened the 'agaverer', was the result.

It was made out of a small spade. To make the blade more rigid, I had the edges bent up at a light engineering works, whose manager was so intrigued by the whole strange business that he did the job free of charge. When an enthusiastic weeder got a bit rough with it and the handle broke, we realised that the little spade was too lightly constructed. It would have been better to have started with a small stout shovel.

When sharp, the 'agaverer' slices leaves sideways with the greatest of ease. Held at an angle and pushed into the soil, the blade picks up and cuts the roots a few at a time and can finally be pushed right underneath the plant to cut the last roots. The notched blade blunts quickly and may need to be sharpened with a file several times in the course of demolishing a really big century plant.

There is plenty of scope for specialised tools like this. You may not need them very often, but they will save you a lot of time and trouble when you do and they may prove unexpectedly versatile. For example, we later found the agaverer very effective for rooting out bananas.

There is no limit to what you may devise, only to what you are prepared to carry.

Looking after your tools

Make them visible

In the bush, small tools are all too easy to lose. Drop one, or merely put it down for a moment and, if it is an inconspicuous colour, it vanishes. Handles dipped in yellow paint show up superbly against the mulch.

Keep them sharp, clean, greased and adjusted

I admit freely that when it comes to tools I am fanatical about efficiency which, for cutting tools, means that I am a sharpness fanatic. Why do otherwise sensible people squander their valuable time and energy gnawing away at hard wood with blunt tools, when a few minutes' work on an oilstone would halve their labour?

I suggest you buy, beg, borrow or steal a medium-grade oilstone and a can of light machine oil or neatsfoot oil and then either teach yourself to sharpen tools from the directions in a book on carpentry or, better still, get a competent woodworker to give you lessons.

Get a tradesman's oilstone, 175 mm to 200 mm long. I greatly prefer the India type, which costs more than the silicon carbide one but gives a better edge and lasts half a lifetime.

Be kind to your tools. Wipe off the sap and grit they inevitably pick up when cutting roots, and keep their working parts lubricated with grease (I use vaseline) or heavy oil. If your loppers or secateurs are sharp, but feel wobbly and are not cutting properly, tighten the appropriate nuts, but not so much that they do not open and shut comfortably.

To sum up, it is well worthwhile to buy good tools, modified if necessary to suit the purposes for which you intend using them. Look after them well and they will save you both time and effort and be a real pleasure to use.

Sharpening knives, loppers and secateurs

Knives should always be kept sharp. There are many effective ways of sharpening them, but one thing you must avoid is making a shoulder on the blade. You can work the blade up and down the oilstone or tradesman's stone with a circular action, then turn the blade over and repeat the action. I heard of one regenerator who steadied the knife with one hand and sharpened it as for a scythe, with a small kitchen whetstone on a handle. At one stage I sharpened mine as I would a chisel or plane, with long diagonal strokes back and forth, again on each side of the blade.

Loppers and secateurs can be touched up with a little pocket stone, but this will sooner or later wear the blade crooked. Both need to be sharpened to the manufacturer's edge. As they are being held together by nuts and bolts, you will need a couple of spanners or a screwdriver to take them apart so that they can be sharpened on the stone and occasionally on the emery wheel.

USE OF POISONS

We regenerate bush by using methods that give us the most effective kill of weeds and the most bountiful growth of natives; that is, by skilful manual weeding. This can be laborious, and we are often asked, especially when we are having to spend a long time extracting a big weed 'Why don't you poison it?' We prefer not to use poisons if we can avoid it, and we certainly condemn their indiscriminate use.

We have four arguments against *indiscriminate* use of poisons.

1. Poisons are not truly selective

They are lethal to weeds and natives alike. If they gain entry to any part of any plant, they will poison that part. If they gain sufficient entry to a plant's growth system – its trunk, its leaves and its roots – they may kill the whole plant.

For a poison to be selective it must act only on plants with certain external characteristics in common, e.g. those with broad leaves but not those with narrow leaves, or those with hairy leaves but not those with shiny leaves. Since these characteristics are common to both weeds and natives, use of sprays is automatically ruled out.

It should be noted that injection of poison into a weed tree can cause damage to nearby natives. Where their roots make contact, poison can transfer from one to another.

2. Poisons can have long-term detrimental effects on the environment

Many modern herbicides are said to be harmless to anything but the plant they are used on and/or to break down very quickly in the soil. I read this and think of the insecticide DDT. When DDT first came on the market, it was said to decompose in one month. Now it is known that the decomposition consists of a minor change only, and the insecticide remains unchanged in the soil and water systems and has spread through many food chains, right across the globe.

The synthetic plant hormone 2,4,5-T was also said to be perfectly safe. It was *not*. The problem is that, in the synthesis of complicated organic molecules, along with the main chemical reaction that yields the compound wanted, there are a number of side-reactions taking place that yield unwanted compounds as contaminants.

In the production of 2,4,5-T, a very dangerous compound called dioxin is formed. My aversion – noted in *Bush Regeneration* in 1971 – to the use of synthetic hormones was based mainly on the report of the births of an unusually large number of deformed babies, following the spraying of 2,4,5-T on a water supply catchment in the United States. So now – following this and many other reports – most dioxin is being removed from commercial 2,4,5-T. Disposal of this waste is still a problem.

On top of all this, at the back of my mind there is always an uneasy question mark. What about the plants and animals I cannot see? I have heard of no research into the effects, if any, that poisons may or may not have on the microscopic life of the soil.

In short, today's safe poisons may in a few years' time prove to be highly dangerous.

3. Poisons are often dangerous to the user

They may not be lethal in the short run but at best they are not very safe preparations to use. Read the labels carefully if you doubt this statement.

4. Poisons do not always work

Unless we can be sure of success, the benefits of poisoning (i.e. saving of time and soil disturbance) are outweighed by the costs.

To date, our experience with poisoning has been far from happy. If a weed is hard to kill mechanically, it is hard to find a reliable way to poison it. In *Bush Regeneration,* I wrote of two small-leaved privets which were poisoned by pouring arsenical weed-killer into auger-holes drilled in their trunks. A month later they looked very dead. Two years later they shot again. One I grubbed out. The other went on sending up shoots from various parts of its anatomy for eight years.

There are other poisons, other ways of using them. People keep telling me they have the answer ... 'Frilling is better than drilling, scraping and painting poison better than either, painting cut stumps is best of all. Use poisons dissolved in water, in distillate, in sump oil or straight from the bottle. Use arsenic, it is better and safer. Anything can be killed by the right kind of ringbarking – no poison needed ...' and so on, ad infinitum.

However, when we remove a plant mechanically, it is dead and it stays dead. A mistaken judgement which leads to regrowth is rare indeed. We move through the bush very carefully, very skilfully. Nevertheless, every metre we travel takes up some time, every step we take causes some damage. Once we have weeded an area and done the necessary follow-up, we do not need to walk all over it every few months – perhaps for eight years, possibly more – hunting for 'poisoned' weeds regrowing in a sea of regenerated natives. Add to this the infuriating factor of not knowing when you can say you have finished the job, and the costs involved, poisoning in every way exceeds its benefits.

Our weeding techniques are perfectly safe and exceedingly reliable; we use and recommend them for the vast majority of the weeds which we encounter. Nevertheless, for some bush weeds, the hand removal of very large specimens is virtually impossible. We have spent many years experimenting with, and documenting, the use of various poisons, without great success. Anyone who comes to us with five years' documentation of the reliable and reasonably safe poisoning of a hard-to-kill species will be very well received.*

* *Editorial Note:* Since Joan wrote this, years of experiments have been carried out with the use of herbicides. We have had some success in safely using glyphosate, thereby easing the worry of large tree removal. It can be selectively used either by injection or by applying it to the cut trunk or roots of weed trees. It can also be sprayed, mainly to control heavy infestations of exotic grasses or bulbs. Take care to avoid overspraying; do not spray on windy days or in or close to watercourses. Wind and water can carry poisons to pristine areas of bush, harming plants and animals.

THE GENTLE ART OF WEEDING

The following techniques, well tested by many workers over many years, are reliable means of getting rid of weeds. Weeding the bush has been called the gentle art, the emphasis being on releasing native plants from the pressure of weed growth rather than on removing weeds.

Trained regenerators at work are a delight to watch. To the novice they seem to have developed a sixth sense, but it is just the result of accumulated experience. It can be like a surgeon at work. Control and skill take over, each stroke, each movement causing the least amount of outside damage, taking the shortest time (by saving motions) and using the least amount of energy. Even after 15 years we are still learning to perfect our techniques. But the rewards are on our side, gratifying and reassuring. Indeed they are therapeutic, not only to the bush but to ourselves as well.

Our weeding techniques minimise soil disturbance, muscular effort and working time. They may appear irritatingly slow at first, but they grow much easier with practice. It will not be too long before you can take pleasure in using them on huge plants which you once thought could be destroyed only by massive tools in big, strong, heavy hands. The efficient use of light tools makes a vast difference to the rate of progress, both in working time and in waiting time. You will not get results unless you buy our whole package and follow our three principles meticulously.

- Work outwards from good areas to bad
- Make minimal disturbance
- Do not overclear

But be warned – and I repeat, in every sort of plant community, wet or dry, rich soil or poor, rainforest, tall timber or heath, these principles remain the same. If you have not understood the plan of work, or have not worked from good areas towards bad, or if you have disturbed the soil unnecessarily or overcleared, then your whole exercise will be to no avail. In fact, it could have quite a detrimental effect and further downgrade the area.

The site, size and species of plant to be removed determine which tools and which techniques you use. You cannot assume that the bigger the weed, the bigger the tools you need to remove it. Force is not necessarily what is needed. Axes can fell forests and loppers can make short work of wood that would tax your strength using knives or secateurs. But there is a lot of variation within and between the weed species in the way roots grow and how hard they are to get out.

Before attempting any work on any weed, whether it ranges in size from the small crofton weed to the huge camphor laurel, it is necessary to remove all the smaller weed growth around it. It is so much easier to do this at this stage rather than later when you may have inadvertently covered the small weeds. At the same time you will be freeing any native seedlings in the area and this will make it easier for you to avoid damaging them underfoot when you tackle the larger weed.

If there are native vines entwined in a large weed, cut the branches on which they are growing and ease them away from the working areas too. Before you start the operation, remember the following:

1. Remove all small weeds
2. Take care of native vines
3. Note the position of all native plants.

You are now ready to tackle the main weed. There are two techniques involved – pulling and/or cutting.

Introduction to weeding techniques

Many people, looking at a large weed tree and its huge canopy, think that it will be an impossible feat to get rid of it. But it is not what is above ground that you have to worry about. The action is all taking place in the soil. The canopies of trees and vines depend for their growth on their supply of food from the roots, and if you deprive the plant of its food source by severing its roots, you will kill the canopy. (This is why in getting rid of weeds, you have to learn to recognise the different root systems and how to deal with them. See Applying The Weeding Techniques, p. 79).

Pulling technique

How you pull will depend on the root system of the weed. This will be discussed in detail in the following pages.

The one thing common to all pulling is that initially it should be tentative, to test how firmly the plant is anchored. Once you have determined this, your movement must be gentle and steady. Avoid the temptation to jerk as this could break the root. You must pull in the direction of the root's growth, which is the line of least resistance. If the root lies in a lateral position, then pull it laterally. If it is growing vertically, then pull vertically, and if the root grows at an angle of 40 degrees, 60 degrees or 80 degrees to the soil surface, you should pull at the same angle.

Often weed roots pass under the roots of native plants. Pulling them is therefore awkward and you are likely to cause damage if you persist.

You should release the weed by cutting its root further out from the butt than where the native root crosses it, so that when you work to remove the remaining portion of the root you can get a good grip on the newly cut end.

Cutting technique

It is always easier to cut with the grain of the wood than attempt to cut straight across. This applies whether you use knives, secateurs or loppers. Sawing straight through roots with a knife using an up and down motion seems to come naturally to many people but it is quite hard work, it takes a long time and therefore disturbs more soil than it should.

This is where the *thrust and lever* technique comes into its own by reducing the job to two easy motions. It can be applied to a wide variety of weeds from privet to asparagus fern and tufted grasses.

'Thrust and lever' technique.

Mastering this technique is therefore well worthwhile. Work, using both hands. Hold the knife firmly in one hand and keep tension on the stem (or root) with the other. Thrust about three-quarters of the blade into the soil, a little to the side of the root, with the back of the knife at an angle of about 45 degrees to the surface. This will bring the tip of the blade underneath the root. Then bring the blade round towards you, with the tip, still in the ground, acting as a fulcrum.

The weight of your arm makes light work of the thrust. The levering action, which comes not from your fingers but from the end of the handle gripped in the palm of your hand, gives you the advantage of the whole length of the knife and as your hand comes down, it also gives you the advantage of the weight of your arm.

When you cut the root, both your hands will sense the release of tension.

Another cutting technique is the *V notch.* You can use secateurs or loppers in the same way as you would an axe to make a V notch, cutting with the grain.

If you are walking through almost clean bush, armed only with knife or secateurs, and come upon an unexpectedly big weed, your choice is between taking a long walk back (and the chance that you will not easily find the weed again) and using a tool that is less than ideal for the job. Walking back will not only take extra time and energy, it will also mean extra trampling of the bush, which is so bad for the plants. So there you are, confronted with a hefty weed and only a knife or secateurs to deal with it. But, all is not lost. By bending the branch and using the V notch technique your light tool becomes a powerful one.

Tracing roots

In removing many of the larger plants, you will usually have to trace out roots in order to remove them. Some will be surface roots; others will have grown deeper.

- If the soil is light and slightly damp, you can often locate roots with your gloved fingertips, running them one on each side of the root. In most cases, however, you will find you have to use the trowel.
- To trace the outline of the root, hold your trowel at a 45 degree angle to the soil and draw it steadily along each side of the root to expose its surface. Sweep the mulch to one side as you work.

Tracing out root.

- With your trowel tilted outwards and sideways, reach underneath the root and scrape away the soil. This will leave the root lying in a little V-shaped trench. When tracing a big root, hold the blade of the trowel with one hand and the handle with the other. Initially, you should scrape from the stem of the plant outwards.
- When the amount of soil scraped away from the root begins to get in your way, reverse direction and scrape back towards the stem.
- Any lateral (or branch) roots, branching off from the exposed root should be traced and pulled out or cut off, the cut ends being left well buried. If you find the root is anchored by several laterals and you haven't time to pull each one out immediately, cut them where they fork from the exposed root and deal with them after you have removed the plant. Take note of their position, or slightly lift the cut ends, so that they can be easily seen later, when you return to trace them out and remove them.
- Eventually, as you trace, you will find the lateral root is thin enough to cut, or buried deeply enough to be pulled out or cut off. If it is still not possible to lift the exposed root out of the trench, it means that you have missed one or more laterals, usually one that is growing vertically downwards from the root you are tracing. This vertical branch root is called a 'sinker'.
- To deal with sinkers, cut the horizontal root a hand's width from the sinker on the stem side and use this as a T-bar handle to pull on vertically. If the sinker will not pull out, cut it 5 cm below the soil surface (see p. 88).
- When you are using your trowel to expose a root, try to keep your movements smooth and steady. If your actions are too jerky, a lot of soil will spill out sideways from the trench. This makes it hard to retrieve when you need it for refilling holes, and it may bury small plants. When you feel the root is almost ready to pull free, back-fill the trench with soil, most of which will then fall back into place as you lift out the root.

Surface roots, having been cut hard against the stem, then need to be traced some distance before you can safely cut them again. If you cut the root too soon, the ends won't be buried deeply enough to prevent them from regrowing, and you will have to trace the root further and make another cut. This means that you will have two sections of root to dispose of and increases the chance that you may accidentally replant one.

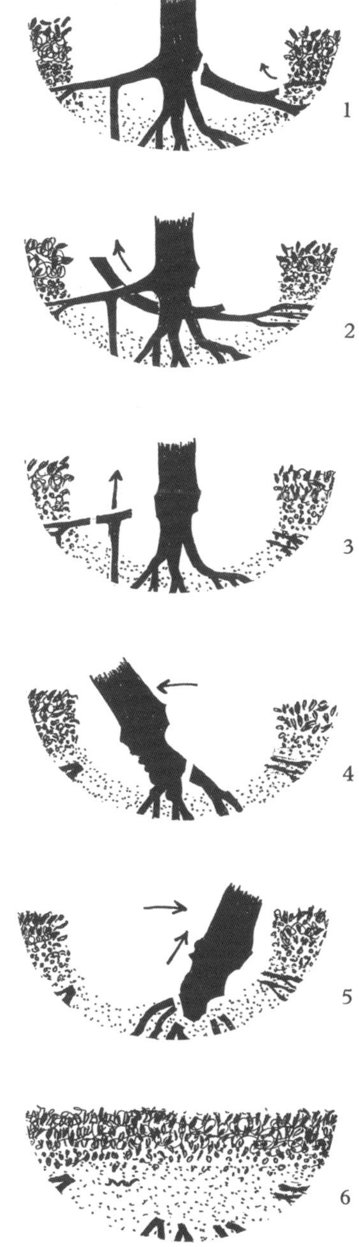

Tracing out roots.

(1) Keep hands low to the ground.

(2) Gently pull the root towards you.

Rocks are a bonus for the bush regenerator. These were hidden under fishbone fern and castor oil plants.

Pampas grass above rock pool.

The area weeded, native plants freed.

Rubbish dumping.

APPLYING THE WEEDING TECHNIQUES

We now come to the application of the basic pulling and cutting techniques. They can be used separately, or in conjunction with one another, on the three main types of plant that you find growing as weeds in the bush – woody plants, vines, and herbaceous plants.

It should be emphasised that weeding conditions vary from one area to another and from one time to another, and that some days you just cannot weed in a particular place. You may be able to pull a weed by hand one season, but have to use a tool on it the next. Soil conditions (moisture, the presence or absence of mulch), plant size and the type of root system you encounter, will all help determine which technique you employ to remove a weed. The methods you will use when disposing of weeds also vary according to the ecology, local conditions and climate.

Before you begin to tackle a large weed, remember to clear the smaller weeds away from round about it, otherwise you may lose sight of them and bury them. Remember also to check whether there are native vines growing in the canopy. If the weed is a privet, lay the canopy on the ground when you have uprooted the plant and the native vine will simply redirect its growth.

Root systems

Most of the woody plants, as well as many of the non-woody, found as weeds in our native bush have a taproot. In this type of system, the

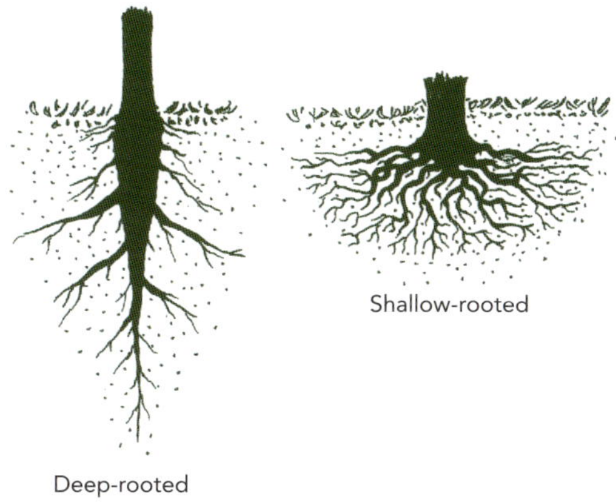

primary or taproot, which is the main root, grows directly downwards from the stem or trunk, and the secondary roots (also called branch roots or laterals) grow out of it. Taproots of the larger plants can grow to a great depth in the soil. The other type of root system is the fibrous root system, in which the roots all radiate out from the stem and no one root is more important than the others. Various grasses have this sort of system.

Many of the weeds we describe have taproot systems with wide-spreading and thickly growing laterals. If the roots are cut deep enough they will not regenerate and can be left in the soil. But if the roots form a mat underground, taking a lot of them out can give the native plants a better chance to thrive.

Woody Plants

Most trees and shrubs are in this category, including camphor laurel, privet and lantana, which all have hard, woody stems. Privet and lantana have a taproot system, with widely spreading laterals.

Shallow-rooted woody plants (e.g. privet and lantana)

Seedlings

REMOVAL

- Seedlings of shallow-rooted shrubs and trees can be pulled up easily with one hand by holding the stem near its base and pulling gently but firmly straight upwards to prevent the plant breaking off short at the roots.
- When the seedling is almost freed, bounce it up and down so that the soil shakes free from the roots and falls back into the hole. With such small weeds, mulch will usually not be disturbed.
- If the mulch is thin or if the soil is held in place largely by mosses and liverworts, this cover may show signs of tearing when you pull on the seedling. In this case, hold the soil down by placing two fingers of one hand close to the stem, one each side of it. As the root comes up soil is stripped from it by the fingers. A knife held flat on the ground can be used instead of fingers to hold the soil down.

- If the plant is growing on a steepish slope, exert the pull in an *uphill* direction, to draw out the roots along their line of growth, which is normally downhill.

DISPOSAL

Place the seedlings you have removed, roots uppermost, off the ground. If you have a great number to dispose of, you can spread them around on nearby rocks or native plants. Make sure that no one native plant is heavily encumbered and that the weeds are not likely to slide off.

Root characteristics vary from species to species. For example, lantana seedlings usually pull out very easily, but you need to grasp them very low on the stem in case it breaks at soil level, leaving you to hunt for the broken end so that you can locate the root. You learn by experience just how much pressure you can exert when pulling on various plants, in order not to break stems or roots.

Fingers help hold soil down.

Saplings

There are two techniques used in removing saplings: pulling and cutting. Although they are described here separately, in practice you will often use both techniques together in removing woody saplings. Pulling is preferable to cutting because it does less damage. Any plant that can be removed by pulling, we refer to as a 'pullable'. A plant may be a pullable when the soil round it is soft after rain, but may not be a pullable during a dry spell.

REMOVAL BY PULLING

- With your hand close to the ground, pull the plant gently, and loosen it by rocking the stem from side to side. Sometimes a gentle jerk will loosen it enough to enable you to pull it out. Only jerk on *strong-stemmed* species, like privet.
- If the plant does not come out, there are probably one or two stubborn roots holding it in. Watch for soil movement when you wiggle the plant, which will show you where these roots are.
- Clear the mulch away along the line of root movement. Stockpile it just far enough away so that it does not impede you. Now do the same with topsoil, keeping it separate from the mulch.
- Remove enough subsoil to expose the root and pull it gently but firmly towards the stem. Keep the hand grasping the root as near the ground as possible and pull parallel to the ground. The root will then draw out lengthwise along its line of growth, where there is least resistance. This is less likely to break it than an angled pull.
- If the plant still refuses to come out of the ground, pull up a few more individual roots until it can be lifted free.
- When the plant has been removed, put the subsoil from around the roots back into the hole, smooth out the loose stockpiled topsoil, firm it down and restore the mulch over the disturbed area.

Privet is a strong-stemmed plant; lantana is weak-stemmed. After a privet sapling has been loosened in the soil, jerking on it gently will often bring it out of the ground, but with this treatment, lantana stems usually break.

If there is no movement of the soil when you wiggle the plant, it means that the roots have turned downwards and grown deeper into the soil. This can be an advantage because, even if they break when you pull, it means they are buried under the soil and will probably not regrow.

If the lantana is growing on rocky ridgetops you need to take care not to break the long roots it sends along the surface or those growing in cracks between rocks.

Whether a plant is pullable or not will depend on its size, its root system and the conditions it is growing in. A plant growing in soft soil can be more easily removed by pulling, than a plant growing in dry, hard soil.

When pulling presents a problem, you will have to resort to cutting to remove saplings.

REMOVAL BY CUTTING

- Rock the plant back and forth to locate the position of the roots by soil movement. Remove the mulch and soil as for 'pulling', to expose the junction of the root with the stem. Cut the root as close to the stem as possible.
- Expose the severed root further until it is at a depth of five times its diameter and cut it again. At this depth, where there is no light and the soil is compact, regrowth is most unlikely. Remove this cut section for disposal.

Push and pull.

Remove small weeds, taking care of any natives.

Stockpile mulch and topsoil. Push stem to one side.
Trace root(s). Cut and draw out roots.

Push stem to other side. Trace root(s).
Cut and draw out.

Lift out, twisting if necessary.

Replace mulch. Dispose of weed.

- Repeat this operation with as many roots as necessary to release the plant.
- Lift the plant out of the ground and shake the soil off the roots, or tap it off with the back of your knife so that it falls into the hole left by the uprooted sapling.
- Smooth and firm the soil and restore the mulch to its original position.

It is sometimes practical to locate roots by prodding downwards near the stem without disturbing the mulch.

When located, the root can then be cut close to the stem. This technique is recommended only if the regenerator has enough experience to know where the root of that particular plant is likely to be.

DISPOSAL

- Hang the smaller saplings carefully on native plants to dry out. Choose natives near at hand.
- Larger saplings can be upended and leant against the trunk of a native plant, or placed on a rock.
- If a lantana has most of its branches already entangled in those of native shrubs or trees, simply bend the plant upwards and hook it by its roots on to the tree it is entangled in.

Note: Place saplings gently on native plants. Do not *throw* them. The force of the throw may damage the natives, and the saplings may fail to stay where they are thrown.

Mature plants

Woody shallow-rooted plants found in the bush can be divided roughly into two groups – those with multiple stems, such as lantana, and those with single stems, such as most privets. In general, the multiple- and single-stemmed weeds have similar root systems. Both types have a root system consisting mostly of lateral roots and it is the root systems that, in the main, determine the difficulty or ease with which the plants can be removed. Other factors may also influence methods of removal. Lantana, for instance, has softer wood than the privet and this can make it easier to deal with than privet, especially when you are cutting roots.

There are two types of privet, the large-leafed privet, with a single stem, which is the easier plant to remove, and the small-leafed privet, which often has several stems, especially if it has been damaged and has regrown.

The canopy of mature plants to some extent determines the method of disposal. Lantana because of its growth pattern is often caught up in branches of native plants and can be left there when severed from the stems. Single-stemmed plants, on the other hand, have to be cut and the canopy removed.

Because the root systems of the multiple- and single-stemmed plants are similar, in general you use the same method to remove them.

Removal of lantana.

Mature plants with multiple stems

The main multiple-stemmed weeds in the Sydney bush are lantana, and those privets that have been damaged and grow multiple stems.

REMOVAL

- Cut a certain number of stems with lopping shears and push the cut stems aside so that you have room to get at the base. Cut several of the strongest stems at a height of about 1 metre to give you leverage on the roots when required.
- Cut the stems, apart from those left as levers, as short as you can

and as close as possible to their base so that they don't impede your movement when you are levering.

- Remove the mulch from around the stump, clearing an area about half a metre in radius. Remove the topsoil from this space and stockpile mulch and topsoil separately.
- To remove the stump, work in pairs if possible, one person applying leverage to hold the big roots taut and the other using the loppers to cut them.
- You can now probably locate the junction of each of the major roots with the stem (or trunk). Cut off these roots as close as possible to the stem. If you cannot see any roots, wiggle the stem and watch for movement of the soil.
- Trace each root outwards from the stem, scraping away the soil, taking care not to mix it with the mulch. Try pulling gently on the root towards the base of the plant to see if it can be drawn out.
- If not, trace the root till it is at a depth of about five times its diameter and cut it again (see p. 88). Pull out the free section of root. Cover the cut ends with soil to prevent regrowth.
- If the plant has a complex root system, you will usually find tiers of roots. If the stump is still held by some deeper roots, probe with your trowel or fingers until you locate them. Then cut them with your knife, secateurs or loppers as appropriate.
- Continue removing roots in this way until the stump is free of the more easily found roots. It is now usually sitting in a small moat and, if it has a simple root system, can at this stage be pulled out.
- If the root system is more complex, continue digging and cutting until you have managed to remove all the deeper roots from all around the stump, except the main vertically-growing one.
- Still working in pairs, loosen the stump by rocking it back and forth, using the 1-metre-high stems as levers.
- If the stump is still held by a deep, vertically-growing root, one person should pull back on the stem levers until the stump lifts high enough to expose the root. The second person cuts the root. This operation may have to be repeated with other vertically-growing roots.
- If one main root anchors the stump, you can try twisting the stump round and round, always in the one direction. This may break the root. It also causes less disturbance to the soil than digging.

Roots are cut at a depth of five times their diameter because we have found that at this depth the cut end of the root does not tend to regrow. You should note that vertically-growing roots have a greater potential for regrowth than those growing horizontally or at an angle. If you are working in an area containing a lot of lantana plants, it would be worthwhile gathering some data on the minimum depth you can cut roots without their regrowing. Try cutting different roots at different depths and observe the results at intervals of about three to twelve months to see which roots have regrown. This will enable you to find the minimum depth necessary to cut to prevent regrowth. Packing soil down on the cut ends with the handle of the trowel compacts it and makes it harder for the root to regenerate.

By making the first cut in the root as close as possible to the stem you will be more easily able to find and to reach other roots, which are almost certainly present at deeper levels in the soil. Also, if you do not cut close to the stem, you will later find yourself driving the cut projection into the soil when you are trying to lever out the root. This is counterproductive!

Working in pairs is an advantage not only when you are trying to

A root cut at a depth of five times its diameter below the soil surface should safely die. Subsoil, topsoil mulch, replaced in that order.

remove large plants, but also when you are dealing with smaller, but tough plants. With one person on each side of the trunk, it can be pushed back and forth to loosen it. This avoids the unnecessary trampling that occurs when one person, working alone, is forced to move round the plants, pushing and pulling to free the roots. With some multiple-stemmed privets it is sometimes necessary to have up to four or five people working on the one plant to remove it.

DISPOSAL

Put the stump on a rock, if possible, or prop it up on a dead log.

Dispose of the canopy of a mature lantana in the same way as you would a younger lantana. The cut stems are usually suspended in the canopy of the native plants. They can be left there, as long as you make sure that all parts are off the ground (see p. 86).

Mature plants with single stems

Most privets come under the heading of shallow-rooted, single-stemmed weeds. With the single-stemmed weeds, dealing with the canopy can be more of a problem than the canopy of multiple-stemmed weeds. Use the same method to remove the roots of these plants as you do for the multiple-stemmed ones.

REMOVAL

- Most single-stemmed plants have the rudiments of a buttress-type root system, with a thickened portion or bulge where the upper roots join the trunk. Cut through the upper roots at the bulge, so that you can get at the lower roots.
- If the mature plant is tall and its canopy is entangled in the branches of other plants, cut the trunk at about waist height by chopping or sawing. The canopy will remain held up in the other plants.
- If the canopy is not entangled, cut the roots and pull on the trunk until it leans. The weight of the canopy will often drag on the trunk and help loosen the main root (taproot) and other supporting roots.
- Work in pairs, one holding the trunk while the other cuts the last supporting roots. Use a rope to guide the plant as it falls, if it is too big to be controlled without.
- There are usually a few unbroken roots, which prevent you from lifting the stump from the hole. Tilt the trunk to enable these roots to be reached.

Cutting the trunk at waist height leaves a long enough portion to act as a lever to loosen roots. If the tree has to be brought down before the canopy is cut, you can cut the trunk to the right length with the tree on the ground. This is easier to do than when the trunk is upright, and you can leave a longer length of trunk to act as a lever. You can put a lot more leverage on a good, strong single stem than you can on the weaker multiples, but don't get carried away by this! If you swing too hard on the stem you can easily break the roots off too short, and the stem can no longer be used as a lever, at least not in the direction of the broken roots.

DISPOSAL OF TREE BUTT

The butt is the cut portion of tree trunk and attached stump.

Butts can usually be left where they fall, unless they are species (e.g. privet, coral tree) that can grow from cuttings. If so, they must be held up off the ground. To do this you can use a rock or a fallen log to hold the butt clear of the ground.

- Use a rock or fallen branches as supports for the butt, near the tree itself.
- Level out soil and restore mulch round and under the supports.
- Work in pairs, one person holding light end of tree, the other swinging the butt round so that it rests on a support.
- Lift light end on to its supports.
- If trunks are light enough to carry, you can lean them, upside down, against the nearest native tree.
- The canopy can be left lying intact where it falls, providing its weight will not damage any natives. If this is likely to happen, lighten the load by cutting off some branches and disposing of them nearby, where they will decay and form mulch.
- Keep all cut ends off the ground or they will strike.

Tall-trunked trees

For a tree that is over seven or eight metres in height you will need to call on the services of a competent axeman, who will make short work of its heavy roots and will control the tree when it falls. You will not encounter such trees very often. Many bush reserves contain no big weed trees at all and even in privet-infested gullies few of these trees reach any great height. You could also consider the use of poison on tall-trunked trees as a last resort.

Deep-rooted shrubs and trees

Camphor laurel, ochna and paddy's lucerne are in this category. The deep-rooted plants have longer-reaching and larger main roots than the shallower-rooted plants, but not such an extensive growth of lateral roots. You must make every effort to remove the large main root completely. If this proves impossible you must cut it very deep down below soil level. Vertically-growing roots store more food than the laterals and have a more tenacious hold on life; they therefore have a great tendency to regrow. This can involve you in much time-wasting work during follow-up.

Seedlings

REMOVAL

- Always pull seedlings of deep-rooted plants straight upwards, holding the stem very close to ground level and using a steady pull.
- If you fail to pull out the seedling by hand, use a trowel. Holding it about 2 cm away from the stem, push the blade vertically down, using the palm of your hand, until the full length of the blade is buried. Move the blade from side to side if the soil is too firm to push the trowel in easily. Now push the handle gently towards the stem of the plant. Do the same on the other side of the stem and try pulling. If the root still does not come out, you may have to repeat the process at various points round the stem until it comes out intact, or you may have to work your fingers down into the space in the soil to grip the root. If you need extra reach, you can use the pliers to pull gently on the root.
- If the soil is too hard, or the plant too big, or the root too kinked, you will have to dig the root out.

Start by sweeping the mulch away from round the stem. Then loosen the soil around the plant by using the trowel as described above. Scoop the soil out with your fingers and put it around the perimeter of the mulch-free area.

Pull on the root carefully until it comes up intact. If it breaks, even deep in the soil where it is very thin, it will regrow.

- Replace the soil and firm it down lightly.
- If you are dealing with a strong wiry seedling like paddy's lucerne,

Other ways of pulling.

which will not pull out when you use finger and thumb, use your hand as a lever in pulling. Put the back of your thumb flat on the ground and clench your fingers round the stem. Roll your wrist over until the little finger bears down on the soil. You should feel the root begin to give and can now remove the seedling with an upward pull. Stop pulling if the root seems likely to break too close to the surface, and use your trowel as described above.

DISPOSAL
Make very sure that all roots are well clear of the ground to prevent them from regrowing.

For the easiest pulling of small woody weeds with long taproots, get to know your soils and watch the weather. There is a certain moisture level at which each type of soil loosens its grip on roots. Days when this happens are redletter days. You find yourself joyfully hand-pulling dozens of plants which you would normally not be able to remove without a trowel. On coarse sandy soil formed from Hawkesbury sandstone this ideal time is shortly after heavy rain, when the soil is drained just enough for you to walk on softly without doing it any harm. On Wianamatta shale, the rich clay loam clings fast when it is thoroughly wet. Here the soil still needs to have had a good soaking, but does not release its grip on roots until it is well on the way to drying out. One of the advantages of our way of working is that we can work in whatever place conditions are right that particular day. So if we find a lot of deep-rooted seedlings like ochna and camphor laurel growing close together, we make careful note of their whereabouts and pull them on a day when the soil conditions are just right. If, however, these weeds are widely scattered they can be easily overlooked and it is best to pull them immediately while you know where they are.

There is a great deal of life in the deeply rooted taproots and if you do not remove the whole of the root, it will send up a new shoot. This can take some time to reach the surface, but it will, and much later you will find it during follow-up. You then have to go through the whole root extraction process again and remove what should have been removed in the first place.

Direct pulling works like a charm on some plants, but if the soil conditions are less than ideal, some species will resist. Ochna roots, for example, develop real kinks. You pull on one seedling and see the stem rising. Although the root seems to be coming out, do not be deceived into pulling harder. The first easy pull has merely straightened some of the kinks in a villainously long taproot which, if broken too high up, will grow again. Camphor laurels have roots shaped rather like parsnips. They also break very easily and without warning. The top of the bulge where stem joins root is a weak point, but the roots can also break lower down.

Saplings

With saplings, which can have very long main taproots, you will have to dig more deeply and make a wider hole than when removing the roots of seedlings.

REMOVAL

- Clear mulch and topsoil from around the base of the stem in a circle with a radius approximately one-third of stem height. Stockpile mulch and topsoil round the perimeter of the circle.
- Dig down close to the stem and all round it. This exposes the root. Put the subsoil you have excavated inside the stockpile of mulch and topsoil, keeping everything separate. The hole can be quite wide, and you continue enlarging it as you dig as deeply as possible, sometimes to arm's length.

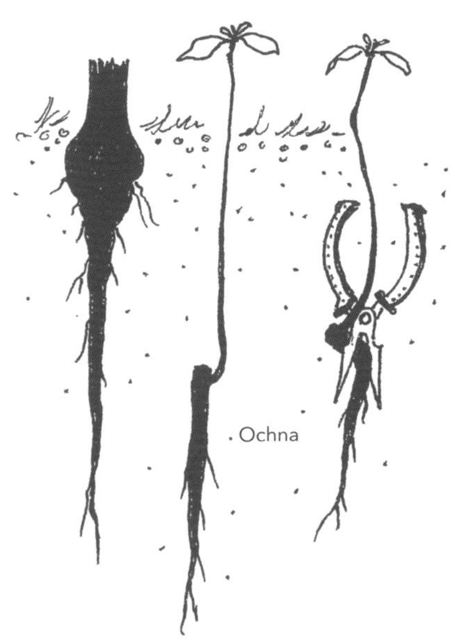

Leave some dead trees for the birds.

Lantana off the ground to dry out.

Kangaroo grass *Themeda australis* – one of the more distinctive native grasses.

The Bradley Bushland Reserve.

- When removing the main root, you must remove the whole of it, otherwise it will reshoot. As the root starts to taper, you can try pulling, or pulling and twisting, to see if it can be drawn out. Pliers can be used to get a grip on the root, but use a steady, even pull. If the root breaks deeply below the surface, where it is pencil thin, ram the soil hard down into the bottom of the hole with the handle of your trowel. This compacts the soil above the cut end.
- Any lateral roots thicker than a matchstick or at a depth of less than 15 cm below the soil surface should be pulled up. Tracing these is usually not necessary; if the hole is wide enough, the laterals will probably have been exposed during digging.
- Replace the subsoil, topsoil and mulch in their original positions.

DISPOSAL

- Leave cut stems and roots well off the ground.

Mature plants

Plants up to two metres in height can be dug out. Those over this height you may have to consider ringbarking or, as a last resort, poisoning. Neither course may be effective and you should make several follow-up visits to see if the poisoned or ringbarked trees show any signs of regenerating.

Ringbarking is clean and easy. This technique involves cutting through the tissues that transport foodstuffs, making your incision right round the trunk or stem. This cuts off the plant's food supply and should kill it. We cherished the illusion that ringbarking was generally effective. Full of confidence, we ringbarked a few lemon-scented gums in Beecroft, Sydney. Lemon-scented gums are native to Western Australia, but can take over the bush on shale in the eastern parts of the country and are counted here as weeds. Not even the tops of our ringbarked trees died. When we cut the ring deeper, some of the tops died, but suckers appeared below the ring and although we repeatedly knocked these off, four years later nearly all the gums were still growing.

We have done little better by using poisons, though we are persevering with experiments (see p. 65) and observing results obtained by other people.

Vegetative reproduction

Many non-woody plants (among which are the vines and herbaceous weeds) can reproduce vegetatively, that is, by some means other than seeds. There are many different ways that plants can do this. Some reproduce by means of runners (or stolons). These are stems that grow along the surface of the soil and send out roots at the nodes. The roots of some plants (e.g. blackberry) can produce suckers, which give rise to new plants, and others, like ochna, and even dandelion, can regrow if a damaged root is not entirely removed. Sometimes leaves are reproductive, as in the mother of millions plants, where numerous plantlets arise along the edges of the leaves.

Tubers, bulbs, corms and rhizomes are all different kinds of fleshy stems, specialised for storage and reproduction. They are usually composed of swollen areas in which the plant stores the food. The tuber is a swollen stem, usually growing underground and with all the typical parts of a normal stem. Some plants, like the madeira vine, can produce both underground and aerial tubers. In the bulb, it is the bases of the leaves that are enlarged for food storage and in corms and rhizomes it is the short underground stem that is thickened. The corm grows upright, often with newer corms attached, and is covered with the dry bases of leaves; the rhizome grows horizontally just below the surface and sends up shoots at intervals. Some plants, especially the grasses, reproduce by means of rhizomes, where each new node can give rise to a new plant. The noxious character of many weeds is due to this type of growth pattern.

Some plants have tuberous roots, that is, the root is thickened for storage (e.g. asparagus fern). These are not, however, true tubers.

The crown is the thickened junction between the stem and the root of a seed plant. The thickened portion stores food and can also reproduce the plant.

Succulents are plants that have tissues specialised for the storage of water and they are therefore adapted to grow where water is in short supply. Examples of succulent weeds found in the bush are the century plant, in which the leaves are succulent, and the wandering jew, in which the runners are the water-storage part.

Vines

Vines are climbing plants with stems too weak or flexible to support them. In the bush they usually support themselves by clinging to other trees and shrubs.

First a warning! Not all vines in the bush are weeds. Some native vines, like devil's twine and kangaroo vine, look as though they are taking over, but they *are* natives and we leave them to their own devices.

Vines can have enormous canopies and be extremely harmful to native plants because of their weight and their tendency to smother.

As with all bush regeneration work, you need to assess your site and decide in what order to remove the weeds. It is generally easiest to take out vines, grasses and scramblers before other types of weeds, but you will find it expedient to remove small weed seedlings before tackling a vine. If you are unfortunate enough to find wandering jew (a succulent weed) and a vine growing together, remove the wandering jew (p. 110) first because it is more fragile and, if broken into pieces, is more difficult to deal with.

Most vines have a simple root system and aerial stems. Others produce runners with nodes, which are the part of a stem at which the leaves are attached. Runners growing on the ground often root at the nodes. Yet other vines use trees and shrubs as their support, but can root at nodes if these are on the ground.

A good place for most vines to die.

Vines with simple root systems (e.g. balloon vine, passionfruit)

Vines with simple roots can be uprooted in the same way as small to medium-sized shrubs.

- Cut the aerial stems. On mature vines, cut them about waist height.
- Pull up the roots of small vines by hand. You may have to sever the roots of larger plants using knife or secateurs, in the same way as you would for the smallish shrubs, and remove them.
- The canopy of the cut vine can be left where it is on its tree support. It will dry in the wind and sun and break down gradually to form part of the mulch below. Removing it would let in a lot more light suddenly, which would stimulate weed growth.

Vines that root at nodes (e.g. honeysuckle and morning glory)

These vines are often found growing near privet or through it, and are much more difficult than privet to remove, because they root from every node. Although the runners of morning glory cover a larger area than those of honeysuckle, it is honeysuckle, with its closely-spaced nodes, which has the greater number of growing points on each runner, and which breaks more easily.

Honeysuckle rooting at nodes (opposite). Morning glory rampant.

REMOVAL

- These vines can often be removed simply by pulling. More deeply rooted specimens may present problems.
- Some vines send out runners under cover of the mulch and low-growing native plants. To remove these runners, work in pairs, one pulling gently on a runner to detect by its movement where it is growing, the other cutting each section of runner, which the first person can then pull out. Where runners have rooted, you can pull out the roots at each node.

- If roots refuse to budge, or are breaking, put slight tension on them by pulling on the attached runner. Then use the thrust and lever technique (p. 69) by placing your knife behind the runner at the node and pulling the knife round towards you, severing the roots well below ground level.
- If a runner breaks, always remove its root, or cut it off below ground level.
- Honeysuckle and morning glory are adept at camouflaging their runners, so it is essential to come back after a period of time to look for signs of green growth among the now brown leaves.

DISPOSAL

- Roll the vine stems up, tie them in a knot and hang them in native trees or shrubs, or put them on rocks. If you like, you can make them into short loops between the natives and use this as a line to hang other light weeds on.
- When the runners have dried out they should be put on the ground as mulch. Don't leave cut runners hanging in the fork of a tree. They can hold moisture which can lead to fungal attack.

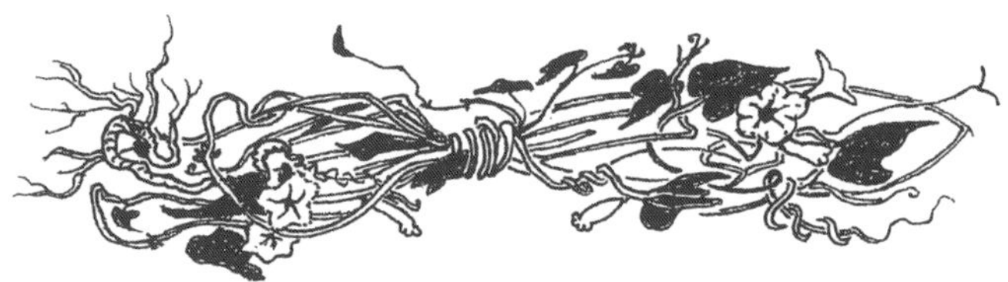

Vines with woody swellings instead of nodes

An example of this type of vine is blackberry, which sends out roots in all directions from woody swellings. It is therefore extremely difficult to remove the entire root system. Blackberry also spreads by means of stolons, which are branches that develop roots wherever the tip bends and touches the ground. Because of this, when disposing of the plants, no portion must be left in contact with the ground.

To remove blackberry, cut the stems and try to remove the root system as you would for a smallish shrub. Frequent follow-up is necessary, to check for regrowth.

If the blackberry is particularly large, careful use of poison may have to be considered.

Vines that root from tubers

Tubers are not always easy to see on the ground, but the ease with which they grow does have one advantage – the new green growth will lead you to them. You must be prepared for frequent follow-up. By monitoring the area you will know how often you must return in order to find and remove any tubers overlooked.

MADEIRA VINE

This vine has both aerial and underground tubers.

- Cut the stems, pull out the roots and put all that you can of leaves, roots and tubers in bags for disposal.

Madeira vine bagged.

What you can't pull gently down, will remain on the supporting plant until the leaves have been drained of food by the aerial tubers. These tubers will finally drop and have to be removed.

- Remove the small ground tubers by holding the stem in one hand and gently teasing the tuber free of the soil with your knife.
- Use a trowel to remove large tubers. Make sure that no piece of tuber is left in the ground, or it will regrow.

POTATO VINE (OR TURKEY RHUBARB)

Potato vine has an extensive system of roots and underground tubers. However, it is the seeds which make this a particularly noxious weed. They are papery light, are borne great distances by the wind, and germinate

Potato vine bagged.

very readily. Ideally, you should remove the flowers in the early flowering stage before the seed is set. Snip the stem bearing the flower very carefully and bag everything – stem, leaves and flowers/seeds.

The potato vine develops tubers of varying sizes and shapes, some like baby carrots, some like potatoes.

- Cut the main stem.
- Put tension on the stem and, using the back of your knife or your fingers, explore in the soil for tubers.
- Dig up *all* tubers and every connecting root, and bag everything for disposal. Restore the surface in the usual way.

Morning glory.

Herbaceous Plants

Herbs (or herbaceous plants) produce, not woody stems, but soft ones. Weeds that fall into this category include: wandering jew, crofton, fleabane, onion weed, catsear, coreopsis, cobbler's peg and dandelion; and the grasses, couch, kikuyu and buffalo.

For the purpose of describing techniques for their removal, we have divided herbs into the following categories: those that root from the base (either shallow-rooted or deep-rooted); those that grow from nodes, crowns, tubers, runners, bulbs or corms; and those that are described as succulents. Many herbaceous plants have more than one method of reproducing. And, of course, most also produce seeds, a further method of spreading.

Removal of seedheads

So far, we have said little about seeds and their removal, apart from mentioning them in connection with potato vine. Some people are horrified when we ignore privets heavy with seed or, at an earlier stage, smothered in flowers. Shouldn't they be cut off, they ask. But unless you intend removing the entire plant within the week, such action, which is in effect pruning, will only strengthen the roots and make the task of eventual removal more difficult.

Seedheads of herbs, particularly annuals, are in a different category from those of trees and shrubs. Many herbs seed freely and the seeds germinate more readily than those of the larger woody plants. Experience will tell you which herbaceous plants in the area you are clearing are likely to pose a problem with seed. Some herbs to watch for and be particularly careful about are crofton, fleabane, onion weed, catsear, coreopsis and pampas.

Always try to remove the weeds before they set seed. If they are in flower when you find them, you must remove the flowers first of all, to ensure that you do not scatter the seeds. If you are too late for this, and if the seeds are windborne ones, put a bag over the seedheads on the leeward side so that the seeds will fall into the bag and you can carry them out. Note the location and make a point of returning at a slightly earlier date the following year to catch the plant before the seed matures and is again scattered.

You may often meet with problems. Imagine that you are just ending a session of work when you come across a weed patch such as fleabane, with the seed-heads about to open, and the weeds growing in a region thick with minute native seedlings where great care is needed. What should you do? Ideally, whenever possible, remove the whole plant immediately. If you cannot do this, then quickly cut off the seedheads and bag them to carry out and dispose of. You must return within the week to remove the plants themselves. Some will have been stimulated to fresh growth and you will find new seedheads are forming and flourishing.

Seedheads in a bag.

Removal of different types of herbs

Herbs that root from the base

SHALLOW-ROOTED (e.g. CROFTON WEED AND COBBLER'S PEG)

These can usually be pulled up by hand or, if they are large, their roots can be cut and the plant removed.

- Try a gentle pull at first to test the soil and the hold the weed has on it.
- Grip the stem with your hand against the ground. This gives good leverage, lessens the likelihood of breaking the stem and reduces soil disturbance. If there is any possibility of damaging native plants, or if the plant refuses to pull out, use your knife or trowel.
- Any soil that comes up with the plant should be shaken into the hole or scraped into it with the back of your knife.
- Restore the soil and mulch as nearly as possible to the original condition.

When pulling larger plants or those with more stubborn roots, such as paddy's lucerne, you can sometimes minimise soil disturbance by placing a foot on either side of the plant as you pull.

(1) Gently push handle against weed, e.g. dandelion.

(2) Do the same on the other side if necessary.

(3) Ease out and up-end on native.

DEEP-ROOTED (e.g. CATSEAR AND DANDELION)

- Push your knife or trowel down into the soil alongside the plant, gently pushing the handle against it.
- With your other hand, hold the plant by its base and gently ease it upwards.
- If there is little or no movement, repeat the operation on the opposite side of the plant (see p. 109).

Dispose of each type of herb by leaving them with their roots clear of the ground. Any that are in flower should be bagged and carried out.

Herbs that root from nodes (e.g. wandering jew – a succulent – and kurnell curse)

Herbs rooting from nodes are prolific growers. Do not confuse wandering jew with the blue-flowered *Commelina cynea,* which is a native. The wandering jew is whiteflowered, but the two are otherwise very similar. Experience will teach you how to differentiate between them.

Wandering jew. Kurnell curse.

REMOVAL

- Keep your hand close to the ground and pull one stem at a time gently and steadily towards you.
- If you find a rooted node that is stubborn, use the back of your knife to free the roots.
- If, with something as prolific as wandering jew, you cannot finish weeding a particular patch, but wish to halt the spread of the herb, lift up the runners and turn them back on themselves (see p. 113).
- At the edge of the bush where you can find heavy infestations of wandering jew, you can roll it up like a carpet. Remember that the stems are weak and break easily, so find all pieces, bag them, and carry them out.

Kurnell curse is aptly named. It is a round, greenleafed runner, introduced to act as a stabiliser on the sand dunes. Any of these plants found in the bush should be removed as soon as possible for, to date, there is no known way of successfully eradicating it.

Herbs that root from the crown (e.g. asparagus fern)

The crown on asparagus fern thickens appreciably and can become quite substantial. It has to be removed and the method we use is called 'crowning'. Asparagus fern also spreads by means of its berries, and although it has tuberous roots, these are only for water storage and do not reproduce.

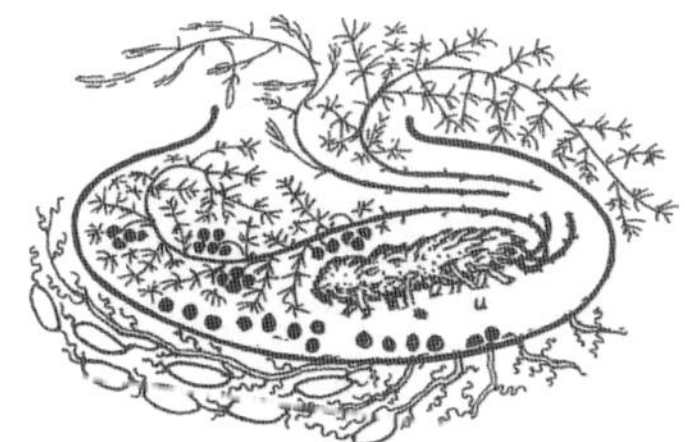

Asparagus fern.

REMOVAL

- In mulch or ideal soil conditions, small seedlings can be pulled gently by hand and then bagged and carried out. If the plants are too big to hand-pull, use the tip of your knife to loosen them in the soil.
- In larger plants, hold the green stems upright and you should find the crown directly below them. Cut off the stems, but leave one or two as markers so that you can find the crown again. These can be surprisingly easy to lose, especially where one overlaps another.
- Scrape away the mulch with the edge of your knife to reveal the edge of the crown.
- Use the thrust and lever technique to sever the crown from its roots. You may have to insert your knife at an angle of 45 degrees at several points around the crown before you can lift it out (see p. 69).
- Large, old crowns can grow as wide as half a metre across. You may find easing up the crown with a jemmy and cutting the roots with a tomahawk the quickest approach.
- Examine the crown to determine whether you have got the whole of it out, or whether a bit has been broken or cut off. If this has happened, find the piece, or it will reshoot.
- Leave in the ground any of the tuberous roots you find. This will minimise soil disturbance and help prevent erosion.

In heavily infested areas of wandering jew particularly on rocks or level ground.

Rolling it up like a rug can be an easy way of removing it.

- If asparagus fern is in flower and you have no time to remove the crown, cut off the flowering stems and do the same with those bearing the green or red berries. Even when separated from the crown and roots, the young berries continue to mature and ripen and, if left, especially on a sloping site, can roll downhill and quickly colonise an enormous area. The berries are also a temptation to birds, who eat the seeds and can distribute them over a wide area.
- Cut stems that have no berries on them may be left on the ground; their leaves provide excellent mulch. Place the crowns where they cannot make contact with the soil, or else bag them and carry them away with you.

Some other herbaceous weeds can also be removed by crowning.

GINGER Ginger has a large and impressive rhizome. This is easy to sever from its roots by crowning. If it is too big to bag, cut it into smaller pieces to carry out.

RIBBON PLANT Its water-storing tuberous roots can be left in the ground to rot, like those of asparagus fern, but you will need to bag the crown and the small plants that develop on its flowering stems.

TUFTED GRASSES such as parramatta, paspalum and pampas, also need only to be crowned.

Pampas grass is becoming increasingly prevalent in the Sydney bush and can grow to a formidable size. Even so, it can be removed by crowning as long as you have carefully cut and bagged any seedheads beforehand. Cut off the leaves and disperse them widely, as they do not mulch easily. Then use your hatchet to crown the plant. Make sure in disposing of it that all pieces of the crown are left clear of the ground.

Pampas.

Paspalum.

Winter grass.

You can work right up to the edges but maintenance will be high.

The complex forest floor; treasure its diversity and tread lightly.

REMOVING PAMPAS GRASS

(1) Remove and bag seedheads.

(2) Use hatchet to crown pampas.

(3) Continuing crowning around plant.

(4) Surprisingly little disturbance.

(5) After careful disposal of leaves and crown restore to area (see page 96).

Herbs that root from tubers

Fishbone fern is an example of a tuberous herb. The wiry roots with their round tubers must all be taken out, and these and the fertile fronds (with the brown spores on the underside) must all be bagged. Remove the root system in the same way as you would a potato vine; fishbone fern is, however, easier to remove. Be careful not to leave any tubers in the ground to regenerate.

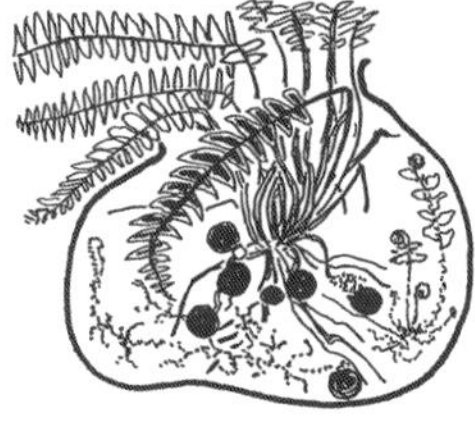

Fishbone fern.

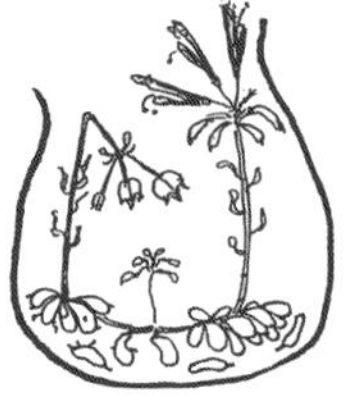

Peruvian Lily

Runners on or below ground (including grasses – buffalo, kikuyu and couch)

Buffalo does not root very deeply. To remove the runners, place your hand close to the ground and, exerting slight tension on the runner, use a sharp knife to cut the roots below each node. The runners can then be carefully hung in a tree or draped over native plants.

Kikuyu and couch grasses are deeper rooted than buffalo and you will probably have to dig down at least 30 cm in order to remove their roots. Be careful to remove them all. Both grasses can be draped to dry out, provided they are securely hung and make no contact with the ground.

Coreopsis is a weed growing from a taproot and producing runners. To deal with this herb, remove the major portion of its taproot and follow the runner to the next plant. Hang the uprooted plant on native plants to dry out, or else bag it.

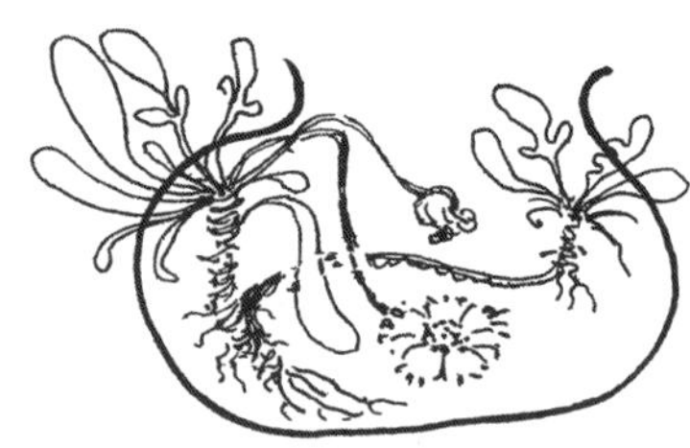

Coreopsis

Buffalo grass.

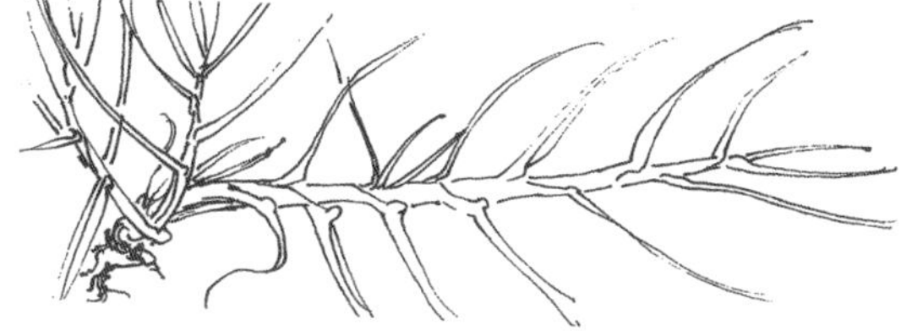

Kikuyu.

Herbs growing from bulbs (e.g. onion weed)

REMOVAL

- Use your trowel to scoop out the bulbs, making sure that all the juvenile bulbs clinging to them are removed also. Even the smallest ones left behind will grow.
- Bag the whole plant to carry out.
- If onion weed has grown so prolifically in the bush that it has formed a mat in which, however, there are few if any native plants, you should consider total removal of all bulb-infested soil.

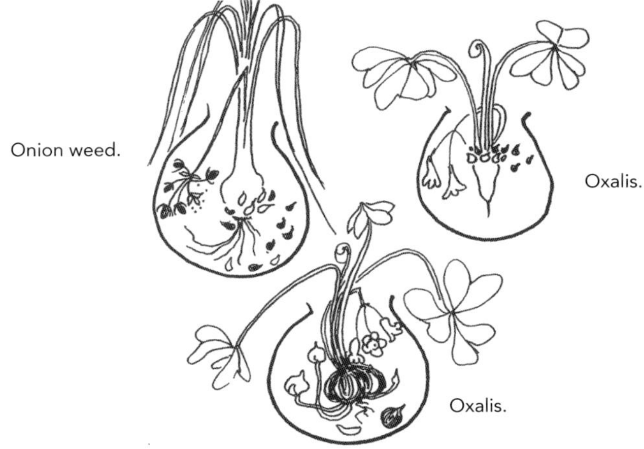

Onion weed.

Oxalis.

Oxalis.

Herbs growing from corms (e.g. gladiolus and watsonia)

Although gladiolus grows from a corm, it is similar to onion weed in that it has numerous progeny clustered around the parent. Remove the corm in the same way as you would a bulb and with just as much care.

Watsonia produces columns of corms in the ground. It also produces cormlets in the axils of the leaves, and these must be removed with as much care as the ground corms.

Watsonia.

Succulents

Wandering jew, a prolifically growing succulent, was dealt with in 'Herbs that root from nodes' (p. 110).

Another succulent that is difficult to deal with is the aptly named 'mother of millions'. You can never be sure you have removed it all. It blends superbly with the soil and plants around it, taking on the colours of its surroundings: grey, pink, a variety of greens and purples. Having found one, you can expect others, even a mat of them, well camouflaged. They are shallow-rooted and must all be pulled out or, before you can blink, some more have grown. Repeated follow-up is necessary.

It is very important to note that all parts of succulent weeds *must* be removed and disposed of, or the portions remaining will regrow.

Mother of millions. Busy lizzie or balsam.

A gladiolus has platelets and cormlets.

Take care to remove them all.

CONCLUSION

Bringing, back the bush is an enjoyable and fulfilling pastime or occupation. The real satisfaction comes later, sometimes with a sudden spurt, sometimes gradually as the natives reassert themselves. The re-emergence of a lost species or the discovery of a new one are added bonuses.

The Bradley method has been successfully tried. You will have unlimited scope in using the techniques and rules outlined here. There will be times when you may have to adapt them to conditions prevailing in your particular area. Different situations could call for different techniques but with experience, common sense and an open mind you will be well equipped to cope, provided that you hold to the three basic principles.

This is not a book about killing weeds. It is a book about growing native plants; not natives bought from a nursery, planted out, watered, mulched, indeed painstakingly and expensively nurtured, but about natives allowed to choose where and when they will grow themselves.

All the insistence on strategy, all the rules, all the tools and techniques, all the preaching of patience, set down here to the point of tedium, are directed to bringing back genuine bush to the natural reserves which have been allowed to deteriorate. As Joan used to say, 'Help the bush to help itself, for the sight of fresh new native plants regenerating of their own accord is a joy for all'.

Following Joan Bradley's death in 1982 a portion of Rawson Park, overlooking Sydney Harbour, was dedicated by Mosman Council as a memorial to the Bradley sisters and named The Bradley Bushland Reserve. Shortly afterwards the Friends of Bradley Bushland Reserve, a voluntary group, was formed. They were successful in attracting a dollar for dollar Bicentennial Grant and a State Initiatives Grant in 1986. The Friends officially opened the Reserve in 1988.

Now the Bradley Bushland Reserve is living proof that the Bradley method is applicable to small pockets of bush as well as to much larger areas. It also shows that it, like many reserves, needs hot fire to prevent it becoming senescent.

The Friends anticipate that the Reserve will remain under their care as a lasting memorial.

GLOSSARY

Exotic Plants (Weeds)

Common Name	**Botanical Name**
Balloon vine	*Cardiospermum grandifiorum*
Balsam	*Impatiens balsamina*
Bitou bush/boneseed	*Chrysanthemoides monilifera*
Blackberry	*Rubus vulgaris*
Boneseed/bitou bush	*Chrysanthemoides monilifera*
Buffalo grass	*Stenotaphrum secundum*
Camphor laurel	*Cinnamomum camphora*
Catsear/flatweed	*Hypochoeris radicata*
Century plant	*Agave americana*
Cobbler's peg	*Bidens pilosa*
Coral tree	*Erythrina* spp.
Coreopsis	*Coreopsis lanceolata*
Couch grass	*Cynodon dactylon*
Crofton weed	*Agerantina adenophora*
Dandelion	*Taraxacum officinale*
Fishbone fern	*Nephrolepis cordifolia*
Flatweed/catsear	*Hypochoeris radicata*
Fleabane	*Conyza* spp.
Ginger	*Hedychium gardneranum*
Honeysuckle	*Lonicera japonica*
Inkweed	*Phytolacca octandra*
Kikuyu grass	*Pennisetum clandestinum*
Kurnell curse	*Hydrocotyle bonairensis*
Lambs' tails/madeira vine	*Anredera cordzfolia*
Lantana	*Lantana camara*
Madeira vine/lambs' tails	*Anredera cordifolia*
Morning glory	*Ipomoea indica*
Mother of millions	*Kalanchoe tubifiora*
Ochna	*Ochna atropurpures*
Onion weed	*Nothoscordum inodorum*
Paddy's lucerne	*Sida rhombzfolia*
Potato vine/turkey rhubarb	*Rumex sagittatus*
Privet – large-leafed	*Ligustrum lucidum*
small-leafed	*Ligustrum sinense*

Common Name	Botanical Name
Ribbon plant	*Chlorophytum comosum*
Turkey rhubarb/potato vine	*Rumex sagittatus*
Wandering jew	*Tradescantia albifiora*
Watsonia	*Watsonia bulbillifera*

Native Plants

Devil's twine	*Cassytha paniculata*
Kangaroo vine	*Cissus antarctica*
'Native wandering jew'	*Commelina cyanea*
Pittosporum	*Pittosporum undulatum*

Healthy bush at Chowder Head.

INDEX